Simple Kitchen Experiments

Learning Science with Everyday Foods

Muriel Mandell

Illustrated by Frances Zweifel

Sterling Publishing Co., Inc. New York

For Horace
who ate the experiments

ACKNOWLEDGMENTS

I would like to acknowledge my debt to food writers Harold McGee, Carol Ann Rintzler, Jane Brody, Bea Lewis, and Howard Hillman. My thanks to biochemist Dr. Betty Rosoff and Dr. Mark Mandell for their technical help, to Jean Mandell and Rose Jutkowitz for their cooking expertise, and to Horace Mandell, Jonathan Mandell, and, of course, my editor at Sterling, Sheila Anne Barry, for their editorial help. And thanks also to Aviva Michaela Mandell, for being.

Library of Congress Cataloging-in-Publication Data

Mandell, Muriel.
Simple kitchen experiments: learning science with everyday foods / Muriel Mandell ; illustrated by Frances Zweifel.
p. cm.
Includes index.
Summary: Includes a variety of simple experiments involving food and cooking principles examining such questions as the effects of heat on different foods, the difference between baking powder and baking soda, and the role of salt.
ISBN 0-8069-8414-7
1. Food—Experiments—Juvenile literature. 2. Cookery—Juvenile literature. [1. Food—Experiments. 2. Experiments. 3. Cookery.] I. Zweifel, Frances W., ill. II. Title.
TX355.M355 1993
641.3—dc20 92-41479
CIP
AC

10 9 8 7 6

First paperback edition published in 1994 by
Sterling Publishing Company, Inc.
387 Park Avenue South, New York, N.Y. 10016

Distributed in Canada by Sterling Publishing
% Canadian Manda Group, P.O. Box 920, Station U
Toronto, Ontario, Canada M8Z 5P9
Distributed in Great Britain and Europe by Cassell PLC
Villiers House, 41/47 Strand, London WC2N 5JE, England
Distributed in Australia by Capricorn Link Ltd.
P.O. Box 665, Lane Cove, NSW 2066
Manufactured in the United States of America

Sterling ISBN 0-8069-8414-7 Trade
0-8069-8415-5 Paper

CONTENTS

BEFORE YOU BEGIN

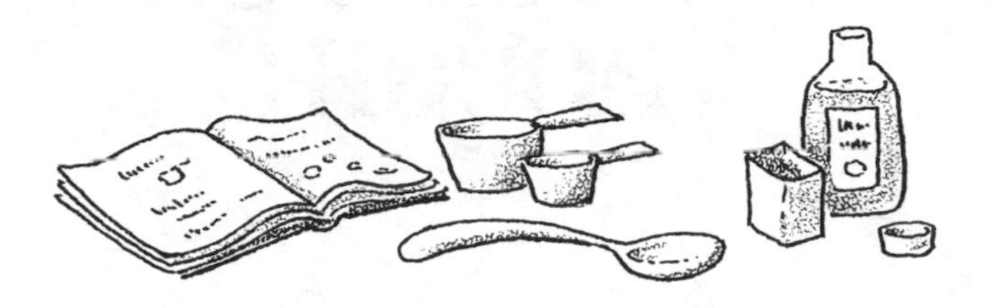

What makes us hungry or thirsty? Does water always boil at the same temperature? Why is a tomato called a fruit? Is a raw carrot healthier than a cooked one? Why do we cry when we peel an onion?

These are a few of the questions the experiments in this book will answer. We'll be working with carbon, hydrogen, oxygen, nitrogen, phosphorus and sulphur—the elements that play a leading role in the chemistry of the kitchen. They combine to make up the food we eat: carbohydrates such as sugars and starches; fats and oils; proteins such as meat and eggs; and water.

When we cook, we're actually preparing chemical compounds in a form our body can use safely—with enough good taste so that we're willing to eat them.

This book will help you discover both the how and the why of doing this. The eating is a bonus!

It is always a good idea to read the "you need" section of an experiment and to gather the equipment and the ingredients listed *before* you start. You may be able to find a substitute for an unavailable item. But you don't want to have to stop in the middle of the experiment. Also read the "what to do" section and follow the directions carefully or you may not get the best results. When you need to use a sharp tool or a stove for an experiment, do get help if that's the rule in your house.

1. FOOD FOR THOUGHT

What makes us hungry? Why do salted potato chips make us thirsty? How does the temperature of food affect its taste? What does salt do in the pot—and in our bodies? Is it better to use honey than sugar? And more. . . .

ABOUT HUNGER AND FOOD

Why do we cook? First of all, to make food digestible and safe. But we also cook for other reasons. We cook to make food tasty so that we will enjoy eating as we satisfy our hunger. But what makes us feel hungry?

Chemicals in the body—in our blood, our digestive hormones and our nervous system—all give us signals, sensations such as stomach movements. When these signals reach the brain, it recognizes them as a need for food.

Our sense of hunger and satisfaction is also influenced by other things. Sometimes we want and eat food when we don't need it. Maybe we eat because it's our favorite food, or because we are upset about something and think food will make us feel better. Or maybe we eat just because everybody else is eating. Sometimes, too, we refuse food even though we need it—perhaps because we are sick or worried or afraid we'll gain weight. Sometimes, we don't eat enough because we don't like the taste or smell or looks of a particular food.

Mapping Your Tongue

Several hundred tiny bumps on the surface and under the tongue help us experience various tastes.

You need:

water
2 tsp salt
2 tsp sugar
aspirin
2 tsp lemon juice
4 small cups or glasses
4 cotton swabs or toothpicks with paper towels
paper towels
paper and pencil

What to do:

Place two ounces of water in each cup.

To #1, add two teaspoons of salt. To #2 add two teaspoons of sugar. Break up an aspirin in a spoon and add it to #3. To #4, add two teaspoons of lemon juice.

Wipe off your tongue with the paper towel or a tissue to get rid of saliva. Dip the cotton swab into #1 (the salty solution). Shake off any drops of liquid and touch the swab to the middle, edges, and back of your tongue. Where was the sensation of saltiness strongest? On which part of your tongue? Write down each answer as you test.

Then rinse your mouth with cold water. Dry your tongue and repeat the process with the three other solutions.

What happens:

You probably sense saltiness and sweetness best at the tip of your tongue. Many people taste bitterness most at the back of the tongue and sourness at the sides.

Why:

The sensation of taste arises from the activity of clusters of cells (the *epitheliads*) that are embedded in the small bumps (the *papillae*) on the tongue's upper surface.

The taste buds in these areas contain nerve endings that respond strongly to each particular taste, and they send their messages on to the brain.

Tasting Through Your Nose

The smell of a food is as important as its taste! In fact, its smell actually influences how it tastes! If you doubt it, try this experiment.

You need:

a small peeled potato
2 spoons
a small peeled apple
a grater

What to do:

Grate part of a peeled potato and put it on a spoon. Grate an equal amount of a peeled apple and put it on a second spoon.

Close your eyes and mix up the spoons so that you're not sure which is which.

Hold your nose and taste each of the foods.

What happens:

You will have trouble telling which is which!

Why:

The nose shares the airway (the *pharynx*) with the mouth. Therefore, we smell and taste food at the same time.

Only salty, sweet, bitter and sour are pure tastes. Other "tastes" are combinations of taste and odor. Without the help of your nose, you may not be able to tell what you are eating.

Some Like It Hot

Believe it or not, the temperature of food affects the taste!

You need:

a glass of cold water with ½ teaspoon of salt
a glass of cold water with ½ teaspoon of sugar
a glass of cold water with the juice of half a lemon
a food thermometer (optional)

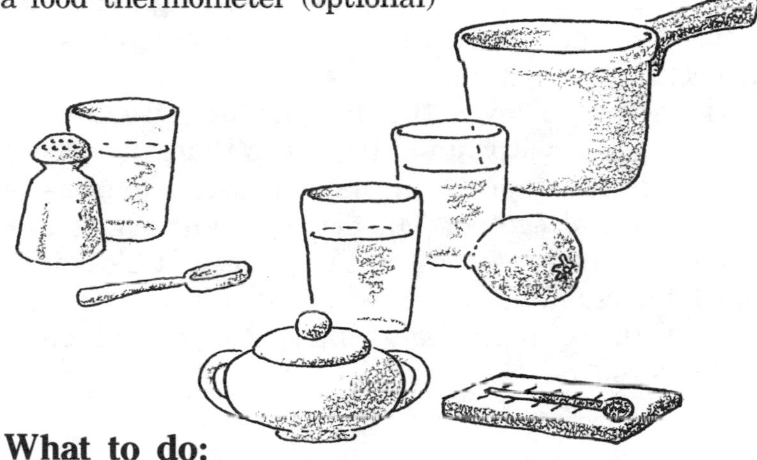

What to do:

Taste the cold salty water. Let it stand on the table for an hour or so until it is at room temperature. Taste it again. Then heat it slightly and taste again. Bring the salty water to a boil. Let it cool slightly and taste it again.

Repeat the process with the sugary water and finally with the lemonade.

What happens:

The salty water tastes saltier at room temperature than at the other temperatures. The sugary water tastes sweetest and the lemonade most sour when they are just slightly warm.

Why:

The temperature at which tastes are strongest ranges from 72° to 105°F (22° to 40°C). Salty and bitter tastes are stronger at the lower range, which is room temperature. Sweet and sour tastes are stronger at the upper part of the range.

If food is *really* hot or cold, it is hard to taste it. The receptor molecules on the tongue can't easily capture the food molecules. Ice cream makers, for instance, have to use twice as much sugar as they would if the ice cream were served at room temperature.

However, whatever the temperature, we are all much more sensitive to bitter tastes than any others.

Certain substances can make a taste stronger—or get rid of it altogether. The cynarin in artichokes, for instance, makes everything taste sweet. It blocks the other tastes. MSG (*monosodium glutamate*), used in some Chinese restaurants, makes salty and bitter tastes stronger.

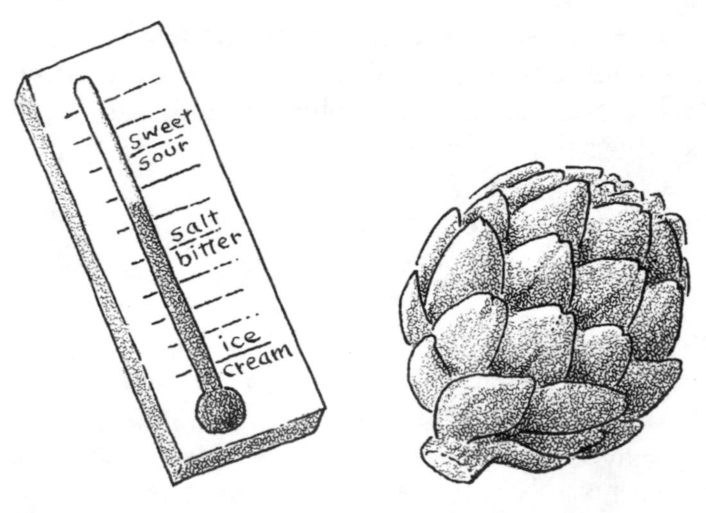

Wilting a Cucumber

Though our bodies need the two minerals (*sodium* and *chloride*) that salt provides, too much salt—and too much of the foods preserved by salt—can cause health problems.

Salt is powerful! Let's look at its effect on a vegetable.

You need:

cucumber or lettuce
salt

What to do:

Cut off two or three slices of cucumber or tear off several leaves of lettuce. Salt them and let them stand.

What happens:

They wilt!

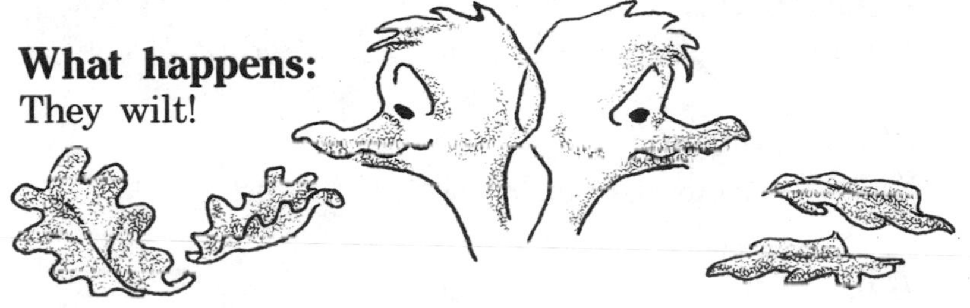

Why:

The salt draws the water out of the cells of the vegetable.

The same thing happens to our body's cells if we eat too much salt and the amount of sodium in the fluid surrounding the cells is too high. The "wilted" cells don't function properly.

Salt is an essential part of blood and other body fluids. But when we eat too much of it, too much water and potassium are drawn from the body's cells. This may cause high blood pressure or kidney damage.

Too Many Potato Chips!

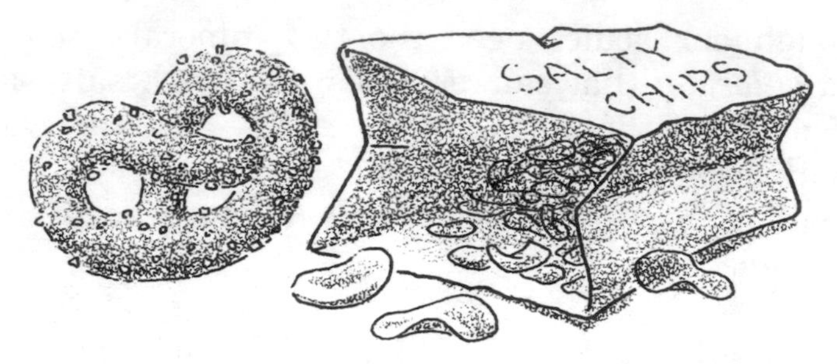

Here is one way the body keeps us from getting too high a concentration of salt.

You need:

a salt pretzel or several potato chips

What to do:

Eat a salt pretzel or three or four salted potato chips.

What happens:

You will need to take a drink of water!

Why:

You want extra water to dilute the extra salt you have eaten. Thirst keeps the body functioning by making sure it doesn't have too high a concentration of salt.

When there is too great a concentration of sodium and potassium in our body fluids, the hypothalamus (a part of the brain located near its base) triggers a feeling of thirst.

An increased concentration of sodium also can be caused when we perspire. You may have heard that you should eat salt when you're sweating, but that's a mistake. Your impulse—to drink water—is right.

WATER, WATER—EVERY TIME

It was once believed that athletes and dancers should never drink water after working out. But today we know it's important for them to drink plenty of water before and after they perform—and even during the competition or game if it lasts long.

Not drinking enough water can cause an athlete to lose the race, miss the hoop, drop the catch. If their muscles don't have enough water, they feel weak and tired.

Although salt is lost along with sweat, the amount is less than the amount of salt in the blood. That means that we need to replace more water than salt when we sweat a lot. Doctors tell us that the safest way to replace lost sweat is with plain water.

Too Salty!

Besides sprinkling on too much salt, what makes food taste too salty?

You need:

salt | water | 2 pots

What to do:

Add two tablespoons of salt to two cups of water. Stir and pour half into one pot and half into the other.

Boil the water in the first pot for 20 minutes. Boil the water in the second pot for 10 minutes. Let them cool and taste each one.

What happens:

The first will taste much saltier than the second.

Why:

After the boiling starts, the water turns into water vapor, an invisible gas, and escapes into the air (*evaporates*). Continued boiling does not raise the temperature. It just speeds up evaporation.

The longer the salted water boils, the more water evaporates and the saltier the remaining water tastes.

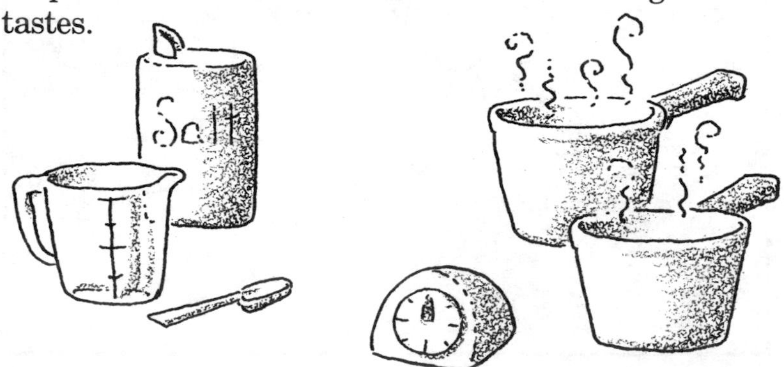

WATCH THAT SALT SHAKER

Food tastes most salty at room temperature. When we season hot food such as potato salad that is going to be eaten cold, we can use less salt, knowing that it will taste saltier when we eat it. When we're salting cold food that is going to be reheated later, we can use a little more salt than our taste tells us—or better still, wait until after we heat it and salt it then.

Which Boils Faster—Salted or Plain Water?

What effect does salt have when you boil water?

You need:

2 small pots half full of cold water

2 T salt

What to do:

Add two tablespoons of salt to one of the pots of cold water. Don't add anything to the other pot. Heat both pots on the stove. Which one starts to boil first?

What happens:

The pot *without* the salt boils first!

Why:

The point at which a substance changes from a liquid to a gas is called the boiling point.

The more salt in water, the higher the temperature must be for the water to boil. Salt molecules turn to gas at much higher temperatures than water molecules.

So we add salt to cold water if we want our food to cook faster since it will be cooking at a higher temperature. Spaghetti and other pastas, for instance, cook well in vigorously boiling salted water. The difference in temperature between salted and unsalted water can be important when we're cooking sauces and custards that call for exact temperature or timing.

What Pot?

Does it matter what size pot we cook in?

You need:

a tall, narrow pot
a short, wide pot
2 cups of water

What to do:

Pour a cup of water into each pot. Place both pots on the stove over a medium flame.

What happens:

The water in the short pot boils first!

Why:

There is less atmosphere in the shallow pot. That means there is less air pressure keeping the molecules down and they have an easier time escaping into the air. The tall, narrow pot is under greater pressure from the air, its molecules have to work harder to escape into the air—and so its boiling point is about 1° higher.

Poached Egg Physics

In which pot—the short, wide one or the tall, narrow one—can we poach an egg faster?

You need:

a tall, narrow pot
4 cups of water
a slotted spoon
a short, wide pot
2 raw eggs
a timer (optional)

What to do:

Pour two cups of cold water into each pot.

Place the short pot on the stove over a medium flame. After the water boils, carefully break open one of the eggs and slip it into the water. Set the timer at two minutes or count 120 seconds. (You do this by saying, "And 1, and 2, and 3," and so on up to 120.) Then quickly remove the egg with a slotted spoon.

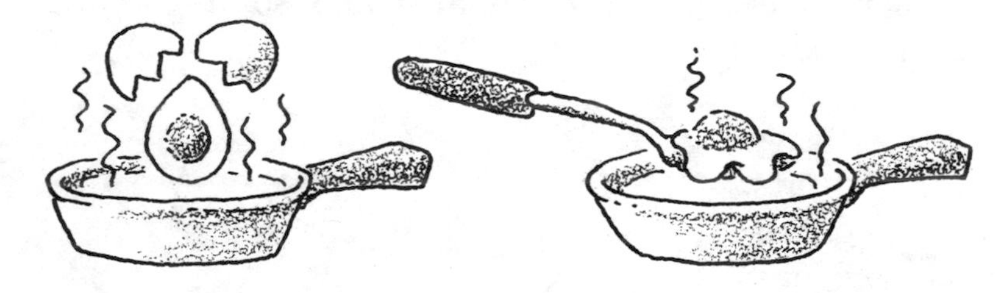

Repeat the process with the egg in the tall pot, again allowing it to cook exactly two minutes.

What happens:

The yolk of the egg in the tall pot gets harder than that of the egg in the short pot!

Why:
Because the boiling point is higher (see page 18) in the tall pot, the food cooks at a higher temperature than in the short, wide pot. Therefore, it takes a shorter time to cook.

POACHED EGGS ON TOAST

2 poached eggs
2 slices of bread

Toast two pieces of bread and top them with the poached eggs. If one of the eggs is cooked harder than you like it, be sure to make the next one in the shorter pot—or rescue it sooner.

WHERE AND WHEN

At sea level, water bubbles and steams at 212°F (100°C). But for every 500 feet (150 m) above sea level, it decreases 1°F, which means food takes longer to cook.

Water boils at slightly lower temperatures on rainy days when the air pressure is low than on clear days when the air pressure is high. So, food takes longer to cook on rainy days than on sunny ones!

Salt Versus the Sweet Stuff

If someone pulled the labels off identical containers of sugar and salt, could you tell which was which? There are ways to tell besides tasting them.

You need:

¼ tsp salt
¼ tsp sugar
2 small saucepans

What to do:

Place the salt in one of the pans and the sugar in the other. Heat each for a few minutes over a low flame.

What happens:

Nothing happens to one. That one is salt. The other melts and gets brown. That one is sugar.

Why:

All sugars are simple carbohydrates. They all contain carbon and hydrogen and oxygen. Heating sugar separates its molecules. At about 360°F (189°C), the sugar breaks down into water (the hydrogen and oxygen) and carbon. The carbon makes the sugar turn brown (caramelizes it). Have you ever toasted marshmallows over an open fire? Then you've seen it happen.

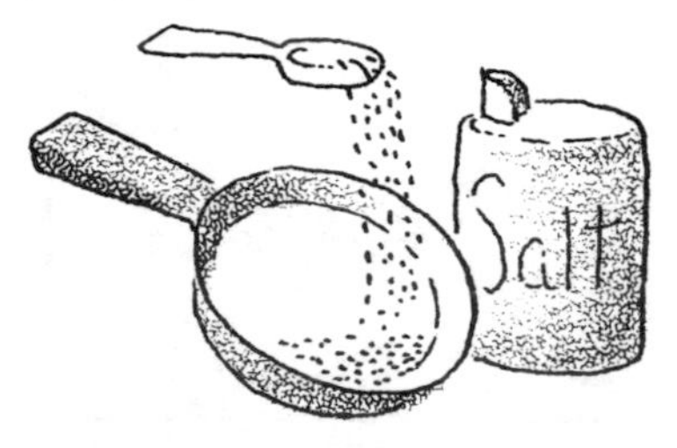

Freezing Salt and Sugar

Here's another way to tell whether a substance is salt or sugar!

You need:
1 T salt
2 cups
food coloring—
  2 colors
1 T sugar
water
an ice cube tray with
  separators

What to do:
Fill the cups halfway with water and color each one with a few drops of a different food coloring. Dissolve the salt in one cup and the sugar in the other.

Pour the solutions into opposite ends of an ice cube tray with separators. Put the tray in the freezer for an hour or two.

What happens:
The sugar cubes freeze. The salt cubes remain liquid.

Why:
Plain water turns into ice at 32°F (0°C). Both sugar and salt lower the freezing point of the water. But sugar molecules are heavier than salt molecules. There are more salt molecules than sugar molecules in a tablespoon. So salt lowers the freezing point twice as much as sugar.

SALT AND ICE CREAM

When you make ice cream, you put milk or cream, sugar, flavoring and gelatin in a special container that sits in a cooling bath of ice-cold water. The water is kept liquid by adding enough salt to lower the temperature to below 27°F (−3°C). That's why salt is such an important ingredient in the making of ice cream.

The Candy Trap

When we feel hungry, we often reach for a candy bar. But suppose we ate a piece of fruit instead?

You need:

1 medium-size apple
1 medium-size banana
Tootsie Rolls
pieces of chocolate

What to do:

For your afternoon snack, try a Tootsie Roll one day and an apple the next. On a third day, try chocolate. On the fourth, eat a banana.

What happens:

They all taste great. But the candy leaves you hungry and wanting more. You may go on to eat three or more Tootsie Rolls and two or more ounces of chocolate. However, chances are that one apple or one banana will leave you feeling full.

Why:

The sugar in candy is highly refined, and it gets digested very quickly. It doesn't stay in the stomach very long and so we stay hungry. Raisins, apples, bananas, pears, and melon contain sugar (fructose) in a form that we digest more slowly, and therefore they fuel the body more gradually.

It's possible to eat a whole pound (.45 kg) of apples before using up the amount of calories in three Tootsie Rolls! Three medium-size bananas are equal in calories to two ounces (56 g) of chocolate.

The piece of fruit also supplies vitamins and minerals and fiber instead of just "empty" calories.

The Cookie Test

Compare the taste and feel of different sugars and honey in these great cookies.

You need:

1½ cups flour (170 g)
1½ sticks or 6 oz. (170 g) margarine or butter
2 T white sugar
2 T brown sugar
1 T honey
1½ tsp lemon juice
bowls
wooden spoon
cookie sheets
teaspoon

What to do:

Preheat the oven to 350°F (175°C).

Soften the margarine at room temperature before you start to mix it with the various sugars.

Using a food processor or a bowl with a wooden spoon, cream a half stick (4 tablespoons) of margarine with the white sugar. Add a half teaspoon of lemon juice. Gradually mix in a half cup of the flour. Continue mixing until the dough is smooth and beginning to form a ball.

Repeat the process with the brown sugar and then with the honey.

Drop rounded teaspoons of the dough onto cookie

sheets about two inches (5 cm) apart. Press each cookie flat with the back of the spoon. Each batch makes about a dozen cookies. Bake 15 minutes or until the cookies are a light brown.

Let cool—and taste.

What happens:

The cookies are equally sweet, but the tastes are different!

Why:

Each sweetener comes from a different source.

White sugar—sucrose—is made either from a tall grass known as sugar cane or from the roots of sugar beets. When it is processed, impurities are removed; it is then refined, and made into the granulated, lump, or powdered form we buy in the supermarket.

Brown sugar also comes from sucrose, but it is made by coating sucrose crystals with molasses, the thick syrup left when water is boiled out of sucrose.

Honey, of course, is manufactured by bees. They make the sweet, sticky thick liquid from the nectar of flowers.

Honey has 18 more calories per tablespoon than sugar, but because honey is sweeter than sugar, you need less. You used only half as much honey as sugar for your cookies.

Honey has small amounts of vitamins and minerals, but too little to offer much nutrition. Cookies made with honey, though, may stay moist longer because honey retains moisture longer while baking. It may even bring moisture from the air into the finished cookies. That's why candy made with honey tends to get sticky.

2. GREEN BROCCOLI AND OTHER VEGETABLES

How do plants eat and drink? When do turnips smell like rotten eggs? How can we make beans user-friendly? The answers to these questions—and much more.

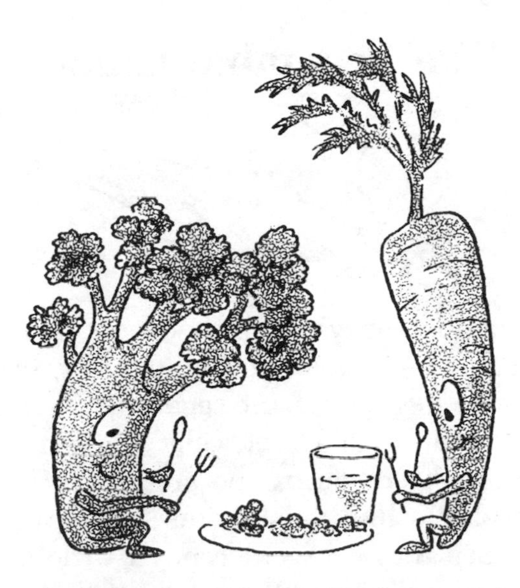

ABOUT VEGETABLES

Vegetables are plants grown for the parts we can eat—root, stem, leaf, flower, seed, or fruit. Sometimes, though, a fruit is a vegetable and sometimes a vegetable is a fruit!

According to botanists, the scientists who study and classify plants, fruits are the part of the plant that contains the seeds. But whether we call a food a fruit or a vegetable seems to depend on its sweetness. Both cantaloupe and squash are in the same fruit family. But squash is not all that sweet and it is served as a vegetable.

Many vegetables are eaten raw. Lettuce and other greens, tomatoes, and cucumbers are among the most usual salad ingredients. Onions and peppers are often added. Cauliflower and broccoli are often eaten raw and dipped in a spicy or creamy sauce. However, many vegetables are easier to digest when they are cooked. Others, like potatoes and yams, can't be digested at all unless they are cooked.

The Vegetable Game

Botanists divide the vegetables we eat into the following groups: leaf, stem, root and tuber, flower and bud, seed and seed pod, fruit-vegetable, and fungi. A carrot, for instance, is a root, celery a stem. Potatoes are tubers, fleshy underground stems bearing a bud. Fungi are plants like mushrooms that live on other plants because they lack chlorophyll—the green coloring matter (see page 32), and so cannot manufacture their own food.

How much do you know about veggies? See if you can identify the following vegetables according to the part of the plant we eat.

a. Root b. Tuber c. Stem d. Leaf e. Flower
f. Seed g. Fruit h. Fungi

1. asparagus
2. beets
3. broccoli
4. brussels sprouts
5. carrots
6. cauliflower
7. cabbage
8. celery
9. corn
10. cucumber
11. eggplant
12. kale
13. leek
14. lettuce
15. morels
16. mushrooms
17. okra
18. onion
19. parsnip
20. peas
21. pepper
22. potato
23. pumpkin
24. radish
25. spinach
26. squash
27. sweet potato
28. tomato
29. turnip
30. water chestnut
31. yam

Answers on page 32.

How to Feed Celery

Plants feed us, but how do plants get fed?

You need:

- a stalk of celery with its leaves
- a half glass of water
- 1 tsp of red food coloring

What to do:

Stand the stalk of celery in a half glass of water colored with a teaspoon of food coloring. Start it off in bright light and let it remain overnight.

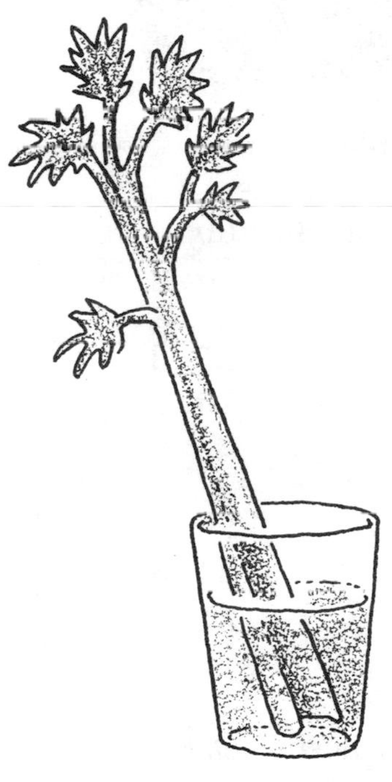

What happens:

The leaves turn reddish.

Why:

The celery stalk is the stem of the celery plant. It absorbs water and minerals from the soil through its root hairs by means of osmosis. Osmosis is a process by which some liquids and gases pass through a membrane—a kind of skin. The water passes into nearby cells and is carried up through its center tubes to the plant's stem and leaves.

The chlorophyll in the leaves—their green coloring—turns the light of the sun into energy. This energy is used to combine some of the water from the soil with carbon dioxide from the air. The carbon and oxygen of the carbon dioxide react with the hydrogen and oxygen of the water to form carbohydrates. This sugar and starch serve as food for the plant—and eventually for us.

You can eat the celery now, or you can use it in the salad on page 39.

Answers to the Vegetable Game

a. Root b. Tuber c. Stem d. Leaf
e. Flower f. Seed g. Fruit h. Fungi

1. c	12. d	23. g
2. a	13. c	24. a
3. e	14. d	25. d
4. d	15. h	26. g
5. a	16. h	27. b
6. e	17. g	28. g
7. d	18. c*	29. a
8. c	19. a	30. b
9. f	20. f	31. b
10. g	21. g	
11. g	22. b	

*bulb—a stem enclosed in fleshy leaves

Storing Carrots

What's the best way to keep carrots, beets and other leafy root vegetables fresh and tasty?

You need:

2 carrots with top leaves
2 carrots with top leaves removed
4 plastic bags big enough to store the carrots

What to do:

Wrap one of the carrots with leaves in a plastic bag that has air holes punched in it. Wrap one carrot without leaves the same way. Store them both in the crisper of the refrigerator.

Wrap the other carrots in plain plastic bags without air holes and store them in the crisper, too.

Observe the carrots daily for a week.

Taste each one of them.

What happens:

The carrot that tastes and looks the best is the one *without leaves* wrapped in the plastic bag with holes in it.

Why:

When the leaves are not removed from the carrots, the sap continues to flow from the root to the leaf, depriving the part we eat of some of its nutrition and flavor. In addition, the leafy tops wilt long before the sturdy roots and start to rot the carrot.

In the bag with holes in it, the air can circulate. This prevents a bitter-tasting compound (*terpenoid*) from forming.

CARROTS HATE FRUIT!

If you like your carrots sweet, it is better not to store them near apples, pears, melons, peaches or avocadoes. All of these fruits manufacture ethylene gas as they ripen. That gas also helps develop terpenoid.

No Way to Treat a Lettuce

People have been eating lettuce since ancient times. There are many kinds—Bibb, Boston, Iceberg, Romaine, Red-leaf, among others. Most of the time, we eat these leaves raw in salads, though we can cook them in a variety of ways—even into soup.

But lettuce has to be treated right!

You need:

2 lettuce leaves
2 bowls

What to do:

Tear one of the leaves into bite-size pieces and place them in a bowl.

Leave the other leaf whole and place it in a bowl.

Let both stand for an hour or so.

What happens:

The torn lettuce turns limp, while the whole leaf stays crisp.

Why:

The torn leaf exposes greater areas to the air, so more of the vegetable's water evaporates and escapes into the air. Tearing also releases an enzyme that destroys vitamin C. That's another reason why it's better not to tear lettuce until just before you serve it.

Salting lettuce in advance also makes it wilt. (See page 13.)

HELPING LETTUCE LAST

To perk up limp lettuce, soak it in cold water. The lettuce absorbs the fresh water and the leaves become crisp again.

Lettuce turns yellow as it gets older because its green chlorophyll fades, allowing its yellow pigments to show through. Lettuce will last for a week in the refrigerator. Don't separate the leaves. Wrap the head of lettuce in damp paper towels and place it in a perforated bag. That way you provide the moisture and air it needs to stay fresh.

Taming an Onion

Cultivated since prehistoric times, the onion has a number of varieties. The most familiar are the yellow and red onions that are very strong and can make people cry if they try to slice them without knowing how to handle them.

You need:

2 onions
running water
a knife

What to do:

Peel both onions. Slice one under water. Slice the other without the water.

What happens:

Your eyes begin to tear when you slice an onion—but not when you do it under running water.

Why:

When you cut an onion, you tear its cell walls and release a gas (*propanethial-sulfur oxide*) that turns into sulfuric acid in the air. Sulfuric acid stings if it gets into your eyes. When you slice onions under running water, you dilute the gas before it can float up into the air.

ONION TALK

Some varieties of mild onion do not irritate your eyes—the Vidalia from Georgia, the Walla Walla from Washington, and the Maui from Hawaii. All of them have a higher sugar content because of the soil and climate in which they grow.

Believe it or not, onions may be good for your heart. A number of laboratory studies find that oils in onions appear to lower blood levels of the "bad" low-density lipoproteins (LDLs), which carry cholesterol into the bloodstream, and they raise the blood levels of the "good" high-density lipoproteins (HDLs), which carry cholesterol out of the body.

CHILL IT!

Another way to tame the onion is to chill it in the refrigerator for an hour or so before you slice it. The cold temperature slows the movement of the atoms in the gas so that they don't float up into the air so quickly.

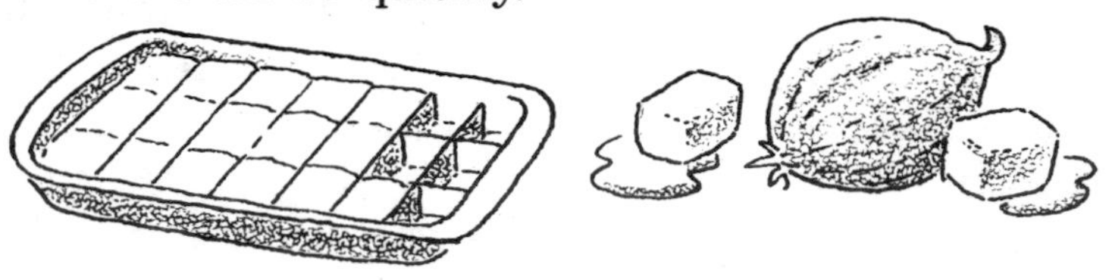

TOSSED SALAD

4–6 lettuce leaves
2–3 onion slices
half a carrot
stalk of celery
½ oz cheese (optional)
a dozen or so croutons
2 tsp oil and 1 tsp vinegar
or 1–2 T of plain yogurt
or 1 T French or Russian salad dressing
salt and pepper

Combine slices of lettuce and onion with whatever your refrigerator holds!

Tear the lettuce into bite-size pieces and cut the onion slices in halves or quarters. Add half a carrot and a stalk of celery cut into small cubes, bits of cheese, and a handful of croutons (buttered toast squares). Stir in either oil and vinegar, or a dollop or two of yogurt, or a tablespoon of French or Russian salad dressing if that's what's on hand. Season the salad with salt and pepper to taste.

Taking the Starch Out of a Potato!

What is starch? It's what people sometimes add when they wash shirts and it is an ingredient in many medicines. But it's also an important food!

Plants make starch from sugar molecules in order to store food for the winter. Plants also use starch to feed seedlings or new sprouts. The starch is stored in the seeds of corn and wheat, in the stem in sorghum (a grain similar to Indian corn), and in the roots or tuber (underground stem) of yams and potatoes.

How do we know potatoes have starch?

You need:

a potato (peeled)
a strainer or cheesecloth
aluminum foil
a paper towel
½ tsp flour
a grater
a bowl
a drop or 2 of tincture of iodine
½ tsp salt

What to do:

Grate a tablespoon or two of potato into a bowl. Squeeze the potato mush through cheesecloth or a fine strainer onto a piece of aluminum foil. Pat the mush dry with a paper towel. Then apply a drop of iodine to it.

Place the salt and the flour on the aluminum foil. Apply a drop of iodine to each.

What happens:

The salt takes on the light brown tint of the iodine. The potato and the flour turn blue-black.

Why:

The blue-black color tells us that starch is present. A chemical change takes place as the iodine combines with the starch. Starch is a carbohydrate, made up of carbon, hydrogen and oxygen.

In the supermarket, you may see packages labeled "potato starch." Inside is a white, powdery substance ground from potatoes by machines. Huge screens filter out the potato fiber, and the potato starch is then left to dry in large vats.

Potato starch is used to thicken sauces and gravy and to replace wheat flour in cakes, if you don't want to eat wheat.

Potato Race

You wouldn't want to eat starchy vegetables such as potatoes and yams unless you cooked them. You need heat to break the cell walls so the potatoes can be digested.

Water boils at 212°F (100°C). Your oven can reach temperatures up to 500°F. Which method of cooking is faster—boiling or baking?

You need:

a pot of boiling water

2 small potatoes of same size

What to do:

Preheat the oven to 450°F (230°C).

Carefully scrub the skin of the two potatoes, but don't peel them. Place one potato on a spoon and lower it carefully into the pot of boiling water. Place the other in the center of your oven.

Using a long fork, test each potato every 10 minutes until it yields to the fork.

What happens:
The boiled potato cooks faster—even though your oven is set at more than twice the temperature of boiling water.

Why:
In both the boiling and baking, molecules of gas or liquid circulate and transfer their heat to the food. But the molecules of bubbling water move more violently than the air currents of the oven. This is because water is much more dense than air (1000 to 1). It delivers heat more efficiently.

TIPS ON SAVING TIME AND VITAMINS

To cook a potato faster, always put it in a preheated oven or in boiling water.

You can also make a potato bake faster by sticking two or three nails in it. They conduct the heat from the stove to the potato.

Some people wrap a potato in aluminum foil, believing it will bake faster. But foil actually slows down the transfer of heat from the oven. Also, because it keeps the moisture from evaporating, it keeps the potato's skin from getting crisp. Aluminum foil is useful, but only to keep the potato warm *after* it is cooked.

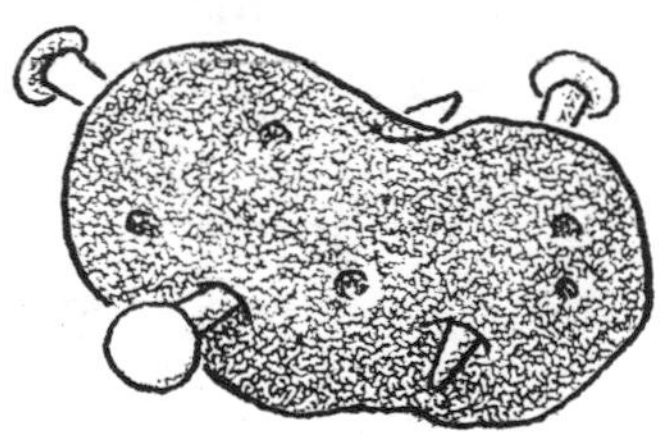

Milking a Potato

Plants draw water from the earth by osmosis. And food gets into our cells by osmosis. We talked a little about osmosis on page 31. What is it and what does it have to do with cooking?

You need:

3 large raw potato cubes of the same size
3 glasses of water
a ruler
salt

What to do:

Put each cube into a glass of water. To glass #1, add a large handful of salt. To glass #2, add a pinch or two of salt. Leave glass #3 plain.

After an hour, measure the potato cubes.

What happens:

#1 will be smaller than it was; #2 will stay same size; #3 will be a little bigger.

Why:

The more salt you add to the water, the stronger the mixture (the solution) becomes. The stronger the solution, the lower its concentration of water.

Osmosis is the flow of a liquid through a membrane (a thin wall). The liquid will always flow into a stronger solution—one where the concentration of liquid is lower.

In #1, the potato cube shrinks as the water in it

moves from the weaker potato juice into the (stronger) heavily salted water. In #2, where the concentrations are equal, there is no movement. In #3, the potato juice is the stronger solution, so the water moves into its tissues and makes the cubes swell.

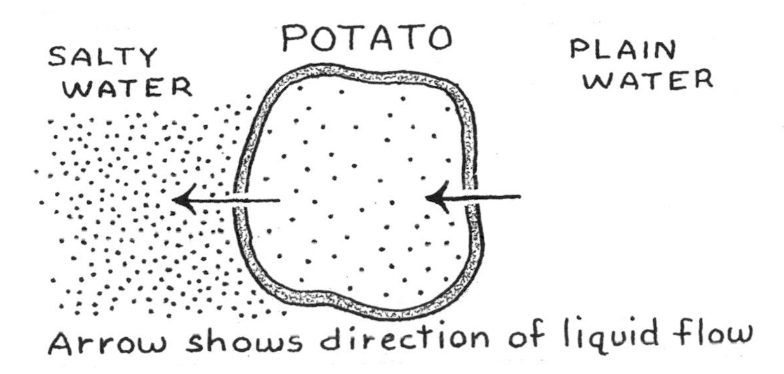

POTATO TREATS

- 2 potatoes from the Potato Race (page 42)
- 1 tsp butter or 1 T milk
- ½ oz (14 g) cheese, sour cream or yogurt
- salt and pepper
- chives or scallions

Mash one of the potatoes you've just cooked with a bit of margarine or butter or milk and season it with salt and pepper. Try the other with melted cheese, sour cream or yogurt. Sprinkle chopped chives or scallions on them for an oniony flavor and a bit of color.

If you cleaned the potatoes well, the skins will be nutritious, too. But the skin of the potato from the oven will be crisper and tastier. The slow baking dehydrates and browns it.

MAKING SOUP

When you make soup, you want the juice to get out of vegetables and meat to flavor the liquid. That's why you add salt to the cooking pot. But when you are cooking meat or chicken as a main course, you want the juice to stay inside. Then you would start with plain water and add salt later, when the cooking is over. Cooking changes the wall of tissue so that osmosis can no longer take place. The liquid can't pass through.

POTATO SOUP

It's easy to make your potato cubes into hot potato soup.

potato cubes cut into quarters
2 cups of water
salt and pepper
1 cup of milk
2 tsp flour
2 tsp butter or margarine
3 or 4 T milk
2 T grated onion
1 or 2 tsp parsley or chives (optional)

Bring the water to a boil, add a pinch of salt and pepper, and put in the potato pieces. In 20 or 25 minutes, when the potatoes are tender, put them in a bowl and mash them. Set aside one cup of the water in which the potatoes were boiled.

Heat the milk until it just begins to bubble. Add the cup of potato water. Stir in the mashed potatoes and continue to heat the soup over a low flame.

In a separate saucepan, melt the butter or margarine. Blend it with the flour and then add the milk. Cook the mixture until it is smooth and bubbly. Then gradually add it to the pot of hot soup, stirring all the while. Add the grated onion. Cover the pot and cook over low heat about 10 minutes. Add salt and pepper to taste. Sprinkle with parsley or chives.

Why Do Some Vegetables Smell Bad?

Some nutritious—and delicious—vegetables don't always get to our plates because of their unpleasant odor!

You need:

a small turnip
a pot of water

What to do:

Peel the turnip and cut it in half. Cut one half into cubes. Leave the other half whole.

Place all the turnip pieces in a pot of boiling water. Test each half with a fork after 15 minutes or so to find out if it is soft. Continue until both halves are firm but tender.

What happens:

The cubed turnip cooks in less than a half hour. The other half needs more time and after a half hour it begins to smell bad.

Why:

Turnips and rutabagas contain hydrogen sulfide, which smells like rotten eggs. When you cook these vegetables, you release this bad-smelling gas. The longer you cook it, the more smelly chemicals are produced and the worse the odor and the stronger the taste.

Shorter cooking time also means that you save more of the vitamins and minerals.

TASTY TURNIPS

You can make both halves of that turnip you cooked taste delicious.

- 2 halves of turnip
- 2 T milk
- ¼ tsp nutmeg
- 1 tsp oil or margarine

Drain off the water. Mash the turnips fine (or puree them in a food processor or blender). Return the turnips to the pot and re-heat them on a low flame, stirring in the oil or margarine, the milk and nutmeg.

OTHER SMELLY VEGGIES

Cabbage and cauliflower also contain smelly sulfur compounds. The longer you cook them, the worse they smell but the better they taste! To lessen the smell, add a slice of bread to the cooking water. And keep a lid on the pot to stop the smelly molecules from floating off into the air.

Keeping of the Green

Green vegetables, such as broccoli, zucchini, spinach, green beans and peas, often come to the dinner table looking drab and unappetizing! Why?

You need:

a few broccoli stalks or
a small zucchini
pot of boiling water

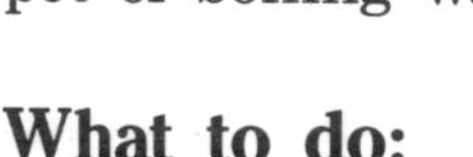

What to do:

Cut off the stem and separate the broccoli flowers. Place the flowers in a pot of boiling water. After 30 seconds, scoop out half the broccoli. Let the rest continue to cook.

What happens:

During the first 30 seconds, the broccoli turns a deep green. The broccoli left in the water loses color.

Why:

The color intensifies because gases trapped in the spaces between cells suddenly expand and escape. Ordinarily, these air pockets dim the green color of the vegetable. But when heat collapses the air pockets, we can see the pigments much more clearly.

Longer cooking, however, results in a chemical change. The chlorophyll pigment that makes vegetables green reacts to acids. Water is naturally a little acid. When we heat broccoli or zucchini or spinach, its chlorophyll reacts with its own acids and the acids in the cooking water to form a new brown substance (*pheophytin*). That's what makes some cooked broccoli an ugly olive green.

Looking Good but Feeling Rotten!

Restaurants sometimes add baking soda to a vegetable to make it look good or cook faster. Does it work?

You need:

broccoli flowers or zucchini strips
a pot of boiling water
1 tsp baking soda

What to do:

Place the broccoli in the boiling water and add the baking soda.

What happens:

The vegetable stays green, but after a short time it turns mushy.

Why:

Baking soda is an alkali, the chemical opposite of an acid. When it is added to the water, it neutralizes some of the acids of the water and the vegetable. Because there are so few acids, the vegetable stays green—but the alkali dissolves its firm cell wall. The vegetable tissue rapidly becomes *too* soft.

Baking soda also destroys the vegetable's vitamins. It's a high price to pay for looking good!

Cold or Hot

We always start cooking green vegetables in boiling water. Why?

You need:

broccoli or spinach

2 pots

What to do:

Place half of the broccoli or spinach in a pot half full of *cold* water. Place the other half in a pot half full of *boiling* water. Cook both until the vegetables are tender.

What happens:

The vegetable in the cold water loses more color than that in the boiling water.

Why:

Plants contain *enzymes*, proteins that cause chemical reactions. They change the plant's color and also destroy its vitamins. The particular enzyme (*chlorophyllase*) involved here is more active between 150° and 170°F (66–77°C) than at other temperatures, so less pigment is lost if the vegetables don't have to be heated through the 150–170°F range. Water boils at 212°F (100°C). If the vegetable is put into boiling water, it avoids the lower range completely.

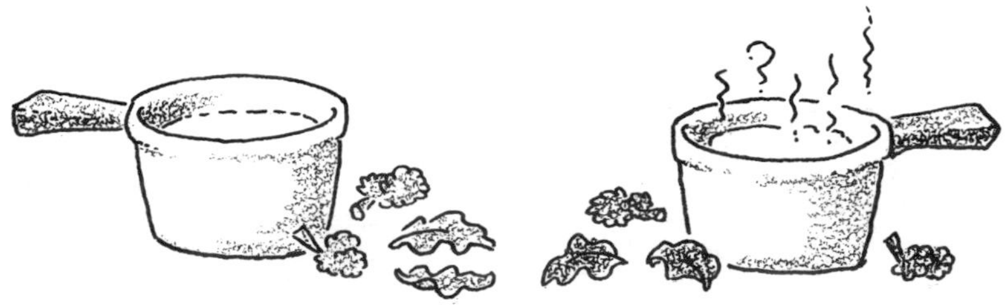

Keeping a Lid On

How can you preserve a vegetable's color better—cooking it covered or uncovered?

You need:

broccoli or zucchini

2 pots, one with a cover

What to do:

Cut off the stalk and separate the flowers of the broccoli or cut the zucchini into quarters.

Cook half the vegetable in a large quantity of boiling water in a covered pot for five to seven minutes.

Cook the other half in a large quantity of boiling water in a pot without a lid for five to seven minutes.

What happens:

The broccoli in the uncovered pot retains its color. The broccoli in the covered pot does not.

Why:

The color changes less in the uncovered pot because some of the plant's acids escape in steam during the first two minutes of boiling. When the pot is covered, the acids turn back into liquid, condense on the lid and fall down into the water.

The bad news is that, without the lid, you lose more vitamins into the air. And because it takes longer to cook without a lid, the nutrients have more time to be drawn out of the food.

There Must Be a Better Way!

In the last experiment, you saw that cooking a green vegetable in a large quantity of water with the lid off prevents it from discoloring. But cooking that way, more vitamins are lost. What's the solution?

You need:

broccoli
a steamer rack (or colander)
a pot with a cover
water

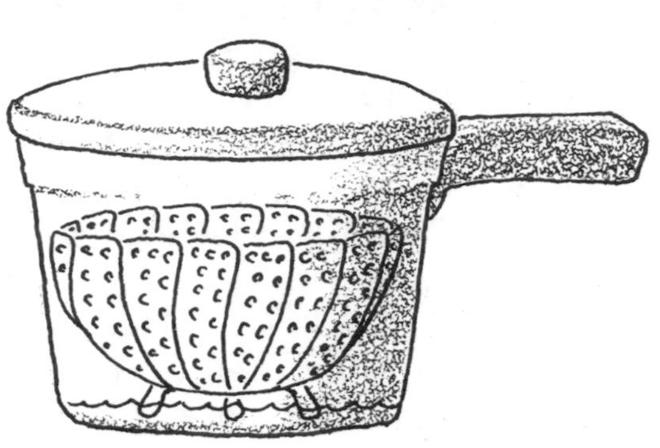

What to do:

Separate the broccoli flowers and put them in the steamer rack.

Place a half-cup of water in the pot and turn on the heat to medium high. When the water starts to boil, slip in the steamer rack with the broccoli, and cover the pot with a tight-fitting lid. Cook for seven to nine minutes.

What happens:

The broccoli remains green.

Why:

Steaming is less effective at conducting heat than water is. So the vegetable may take a few more minutes to soften than it would if boiled. But since it never comes into contact with the acids in the water, it doesn't lose its color or its vitamins.

Colorful Carrot

Unlike green and red vegetables, carrots do not lose their color when cooked in water!

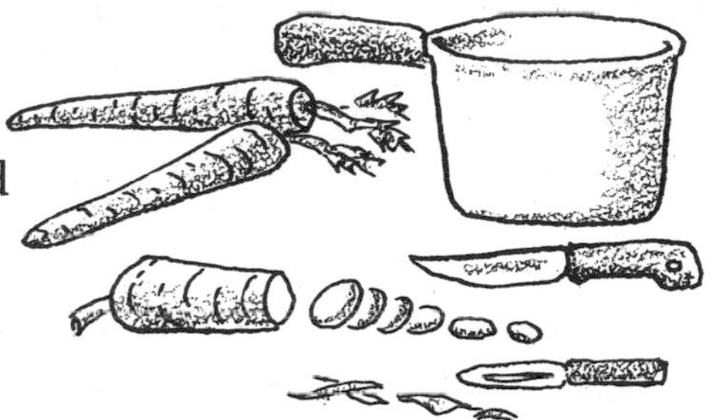

You need:

1 or 2 carrots,
peeled and sliced
pot of water

What to do:

Bring a cup of water to the boiling point.

With a large spoon, lower the carrot slices into the pot of boiling water. Cook for 15 to 20 minutes.

What happens:

The carrots remain orange, but you now can pierce them with a fork.

Why:

Though they dissolve in fat, the coloring matter of carrots (*carotenes*) do not dissolve in water and are not affected by the normal heat of cooking. The carrots therefore stay orange.

Because heat dissolves some of the fiber (the *hemicellulose*) in the stiff carrot walls, cooked carrots are easier to digest and more nutritious than raw carrots. Cooked or raw, carrots are a good source of both vitamin A and the mineral potassium.

Strangely enough, eating huge amounts of carrots (two cups a day for several months) turns your skin yellow! Fortunately, when you stop eating so many carrots, it returns to its normal color.

ABOUT LEGUMES

Legumes are foods such as beans, peas and lentils that come from the fruit or seeds of plants that have pods. We can eat them both fresh and dried. Kidney beans, for instance, are the dried seeds of green beans. But we call the green beans a vegetable and the kidney bean a legume!

Legumes are second in importance to grain as a source of food. They are valuable because they absorb nitrogen from the air. Nitrogen is the essential mineral of amino acids, the building blocks of the protein that helps grow and repair our body's tissues. So, in addition to fibre and essential minerals and vitamins, beans supply us with protein.

Among the thousands of beans that belong to the legume family, only 22 are grown in quantity for us to eat. These range from lima beans, split-peas and chick-peas to adzuki, used in Asian sweet dishes, and the soybeans from which soy sauce and tofu are made.

Though we call it a nut, the peanut is also a legume.

Making Beans User-Friendly

Despite their importance in our diet, beans often make us feel bloated and give us gas. Here's how to make them "user friendly" when we eat them.

You need:

6 T black beans or chick-peas
a colander or strainer
3 pots
cold water

What to do:

1. Place one third of the beans in one of the pots and cover them with a cup of water. Let them stand for an hour. Then, using a strainer or colander, change the water. Place the beans in the refrigerator for four to six hours (or overnight if that is more convenient). Drain off the water and cover the beans with two cups of fresh water.
2. In a second pot, take another third of the beans, cover them with a cup of water. Bring them to a boil and cook for 2 minutes. Then change the water and let them stand, covered, for an hour. Drain off the water and cover the beans with two cups of fresh water.
3. In a third pot, cover the last third of the beans with two cups of water.

Heat all three pots, bringing the water to a boil. Then lower the heat and simmer until the beans are soft. Note how much time it takes for each pot of beans to soften.

Add salt and pepper and minced garlic or onion, lemon juice, dill or parsley. Eat each batch of beans at a different meal and observe how your body reacts.

What happens:

The pre-soaked black beans (either #1 and #2) each take about 1½ hours to soften. The beans that haven't been soaked (#3) take much longer to cook. But even more important, the unsoaked beans make us feel bloated and uncomfortable and give us gas.

Why:

Either the long-term soaking (#1) or the quick-soaking (#2) reduces the cooking time by half because it returns moisture to the beans and softens them. (The shorter cooking time also saves minerals, vitamins and protein that are destroyed during the longer heating.)

Soaking also breaks down the beans' complex sugars, its *oligosaccharides*, which our digestive enzymes cannot digest. When we change the soaking water, we discard these sugars. Otherwise, they ferment in our intestines and produce those bloating and unpleasant gases, mostly carbon dioxide.

Tough Cook, Tender Beans

Most legumes have a bland flavor. We season and flavor them to make them taste good. Mexicans add garlic and chili; Italians add garlic and oregano. The British add mustard and bay leaf. New Englanders make their famous baked beans dish with brown sugar and molasses.

Either tomatoes or lemon juice will make legumes tasty. And their vitamin C makes the iron in the beans easier to absorb. But *not* if our timing is off!

You need:

6 T lentils or
presoaked beans
4 T tomato sauce or
lemon juice
2 small pots
water

What to do:

Cull the lentils (see page 60) and discard any that float to the top. Place three tablespoons in each pot and cover with water.

To pot #1, add the tomato sauce or lemon juice.

Don't add any juice or sauce to pot #2.

Simmer both pots on a low flame. After 20 minutes test the lentils in both pots for softness using a fork. Retest every 10 minutes. Remove the pot from the heat when the lentils are tender but firm. Note how much time it takes them to cook.

What happens:

The plain lentils take 30 to 40 minutes to soften.

The lentils with the tomato sauce take much more time to soften.

Why:

When you add tomatoes or tomato sauce or lemon juice before the beans are soft, the acid of the fruit or vegetable reacts with the starch of the beans to toughen their seed coat. They do soften, but it takes them much longer. That's why acidic foods need to be added only after the beans are soft.

Adding molasses before the beans have softened also interferes with the softening process because molasses contains calcium.

Salt also reacts with the seed coating to form a barrier that keeps liquid from being absorbed and makes the skins tough. It too should be added only after the beans are cooked.

CULLING

We cull beans to pick out small stones and other unwanted material. There is a quick, foolproof way to do it. Place the lentils or beans in a clear jar or bowl. Cover them with a half glass of water. Most of the beans will stay at the bottom of the jar but the few defective ones will float up to the top—they are hollow and therefore lighter. You can lift them out with a slotted spoon.

LENTIL SNACK

½ cup cooked lentils
1 T chopped onion
(optional)

1 to 2 tsp (5–10 ml)
lemon juice
or 1 tsp (5 ml)
tomato sauce

Dried lentils will swell up to two or three times their original size in the cooking. Add lemon juice to them or tomato sauce or vinegar and onions and serve your cooked lentils on crackers. They are even tastier served cold the next day.

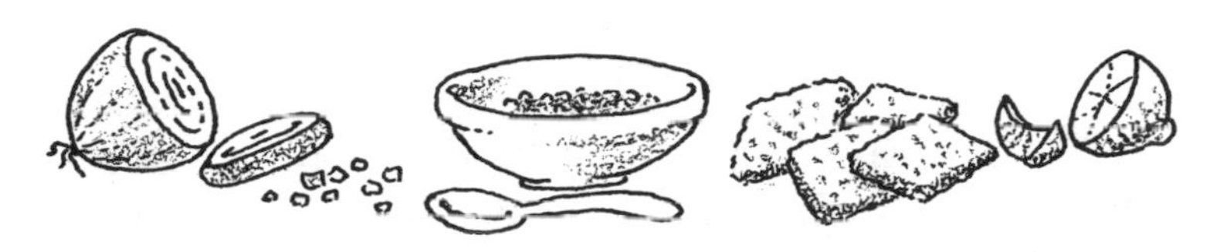

FABULOUS SOYBEANS

Soybeans are the only beans whose proteins are considered "complete." They contain all the amino acids that are essential for us to get through our food.

Soybeans are considered a perfect meat substitute and are ideal for vegetarian cooking. Tofu, a cheeselike curd made from soybeans, is a Japanese favorite that has become popular in the rest of the world. The secret of tofu is that it has no flavor of its own but takes on the taste and smell of the foods with which it is cooked. It can taste like eggs, cheese, chicken—even like chocolate pudding!

Sprouting Beans

The sprouts used in salads actually sprout from beans! You can easily sprout your own.

You need:

6 T mung beans, chick-peas or lentils
cheesecloth
string or rubber bands
labels and pencil
3 clean glasses
warm water

What to do:

Cull the beans.

Soak 4 tablespoons of the beans overnight in warm water. Then drain them and divide them between two clean glasses. Put a layer of cheesecloth over the top of each of the glasses and tie it on with a string or rubber band.

Store the first glass in a warm, dark place like a cupboard. Store the second in the refrigerator.

Without soaking them, place the last two table-spoons of the beans in a third glass. Label it and store it in the cupboard.

Keep the sprouts moist in all three glasses by rinsing them twice a day with lukewarm water. Drain off the excess water through the cheesecloth to prevent rotting or molding.

Note what happens after four or five days.

What happens:
The batch of soaked beans stored in a warm, dark place sprouts and yields four to six ounces (112–168 g) of sprouts!

The beans stored in the refrigerator and the unsoaked beans never sprout.

Why:
During the overnight soaking, the hard outer shell splits, the starch in the beans absorbs the water and the beans swell. That permits the embryo within each bean to take in water. But in addition to moisture, seeds need warmth and darkness in order to grow.

Harvest your crop and use the sprouts raw in salads or on sandwiches, or stir-fry them with scallions, soy sauce and garlic.

ABOUT SPROUTS

As the beans sprout, the starches and sugars in them convert into the energy needed to grow. That's why sprouts have less carbohydrates and are less caloric than beans. In addition, the seedlings produce ascorbic acid (vitamin C). They have three to five times as much vitamin C as the beans from which they sprouted!

Unlike the beans from which they come, sprouts don't create gas in our insides. The complex sugars (*oligosaccharides*) that cause gas are among the carbohydrates that break down to provide energy for the growing sprouts.

3. FRUIT OF THE VINE AND OTHER PLACES

Who calls a tomato a fruit? Why not store bananas in the refrigerator? What does pineapple do to gelatin? Where does vinegar come from? Is one end of a fruit sweeter than the other? And more. . . .

ABOUT FRUIT

Botanists call tomatoes, eggplants, cucumbers and pumpkins fruits because they are the part of the plant that has the seeds.

But as cooks—and eaters—we call them vegetables and save the term fruit for plants that are sweeter.

All fruits grow above ground. We harvest grapes, berries and melons from vines and shrubs. From trees, we get apples and pears, citrus fruits, bananas, figs and dates and cherries. The largest fruit to grow on a tree is the jakefruit, native to southeast Asia. It can weigh as much as 80 pounds!

Most fruits are eaten raw once they ripen. But many are also cooked—stewed, poached, baked—as well as dried, canned, frozen, squeezed into juice, baked in pies, used to flavor other foods, and made into jams and jellies. Some, like plantains, quince, rhubarb and sour cherries, *must* be cooked.

Bite or Bake?

There are hundreds of varieties of apples to choose from. Which do you bite and which do you bake?

You need:

1 Red or Golden Delicious apple
1 Rome, York Imperial, Stayman, Winesap or Jonathan apple
2 to 3 T sugar or raisins

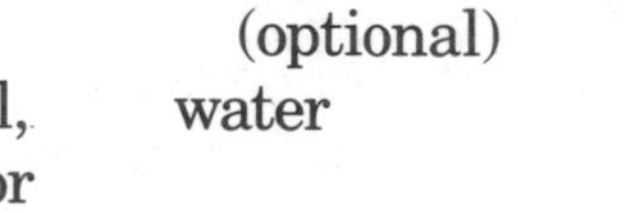

dash of nutmeg (optional)
water

What to do:

Core both apples and cut away a circle of peel at the top. Place them in a baking dish. Fill the center hole of the apples with sugar or raisins. Sprinkle with nutmeg. Add water to cover the bottom of the dish. Place it in a 400°F (200°C) oven for about an hour or until the apples are tender. Taste each one.

What happens:

The Delicious apple is mushy and shapeless.
The Rome is firm and tasty.

Why:

The Delicious apple becomes mushy for two reasons. First, it lacks enough fiber (cellulose), the part of the cell wall that keeps it firm, to hold the peel intact.

Second, the Delicious apple doesn't have enough acid to counterbalance the added sugar. The apple that is less sweet remains firmer and retains more fiber. Fiber is undigestible roughage that is good for us because it helps the intestines and bowels to work better to eliminate waste products.

Of course, a raw apple contains the most fiber!

WHY ARE GREEN APPLES SOUR?

What makes unripe apples sour? Malic acid. All apples have it, but as an apple ripens on the tree, the amount of malic acid declines and the apple becomes sweeter. Depending on the soil and climate in which they are grown, some varieties stay more tart than others. Some people prefer apples like Granny Smiths, which stay green, just because they are sour.

Bursting an Apple

Suppose you *want* a mushy apple!

You need:

2 apples
water
parer (optional)
knife
pot with cover

¼ cup water
1 tsp lemon juice
  or
dash of cinnamon and
  nutmeg

What to do:

Wash both apples, peel them and cut them in four sections. Cut away the core and slice each quarter into cubes. Cook the pieces in a small amount of water in a covered pot until they are tender. Add the cinnamon and nutmeg—or the lemon juice—and cook a few minutes longer.

What happens:

You have applesauce.

Why:

With the peel removed, the pectin—the cementing material between cells that stiffens the fruit—dissolves. The water inside the apple's cells swells, bursts the cell walls, and the fruit's flesh softens. An apple turns into applesauce.

Apple in the Cookie Jar

An apple in the breadbox or cookie jar will affect our bread or cake!

You need:

2 cookie jars or tins
1 slice of bread
2 slices of apple
1 slice of cake (or a cookie)

What to do:

Place one apple slice in a cookie jar with the slice of bread. Place the other in the cookie jar with the slice of cake. Don't open the jars for a day or so.

What happens:

The bread gets stale—and the cake stays moist!

Why:

Sugar dissolves in water. It will absorb water from the atmosphere, if given the chance.

The more sugary food draws water molecules from the other food. The apple has more sugar than the bread, so the bread loses water to the apple. But the cake has more sugar than the apple, so the apple loses water to the cake.

One End Is Sweeter!

Did you know that different parts of the same fruit taste different?

You need:

an orange (preferably navel)
a knife

What to do:

Peel the orange. Cut one slice across the stem end and then one across the blossom end. Taste them.

What happens:

The slice at the blossom end is sweeter.

Why:

The blossom end develops more sugar because it is more exposed to the sun. For the same reason, fruits grown in the temperate zone are only 10 to 15 percent sugar while those from the tropics, such as bananas, figs and dates, range from 20 to 60 percent sugar.

ORANGE AND ONION SALAD

You can continue slicing your orange and pop the slices into your mouth. Or make a Sicilian salad.

6 orange slices ¼″ thick	1 or 2 tsp olive oil
6 red onion slices ⅛″ thick	ground black pepper

Place the orange slices on a salad plate and layer slices of red onion on them. Dribble on a teaspoon or two of olive oil. Grind on fresh black pepper to taste and serve.

How to Ripen a Fruit

All too often, the fruit we buy is not quite ripe. What do we do with it?

You need:

2 unripe peaches, nectarines or other fruit

a brown paper bag

your refrigerator

What to do:

Place one of the unripe fruits in the crisper of the refrigerator for a day or two.

Place the other in the paper bag and close it securely. Put it somewhere out of the way—on top of the refrigerator, for instance. Let it stand for a day or two.

Taste both.

What happens:

The fruit in the refrigerator softens—but it is not very tasty.

The fruit in the paper bag softens—and sweetens!

Why:

In the paper bag, you are trapping and concentrating the ethylene gas that comes from the fruit naturally. This gas speeds up the ripening process. In the refrigerator, the ethylene gas is shared with the other contents of the crisper.

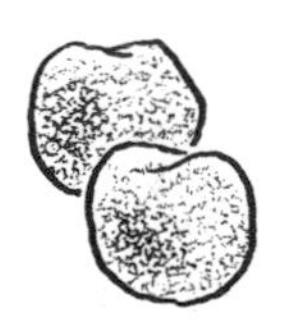

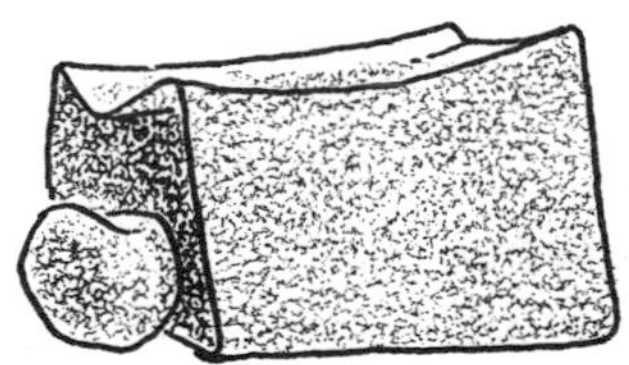

Getting Juice from a Lemon

How do you get juice out of a lemon?

You need:

2 lemons

a knife

What to do:

Cut the first lemon in half and squeeze out as much of the juice as you can.

Before you cut it, roll the other lemon on a hard surface like a countertop. Then squeeze out the juice.

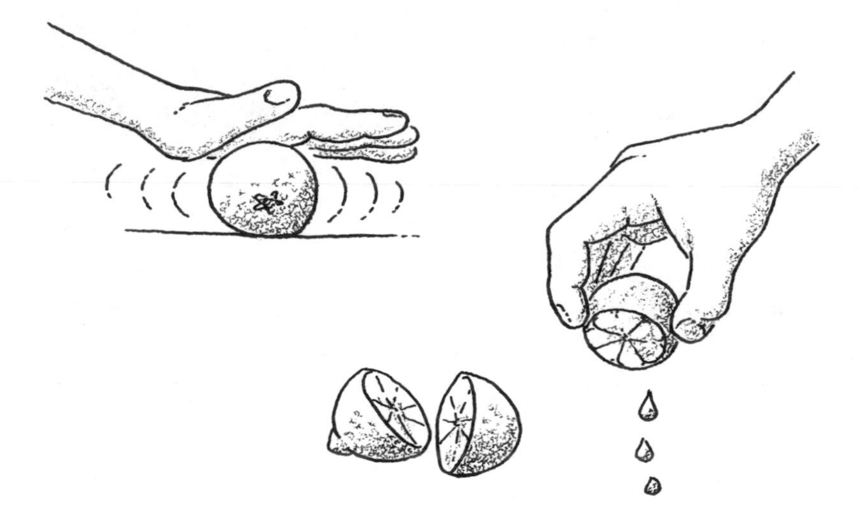

What happens:

It is much easier to squeeze out the juice after you've rolled the lemon—and you wind up with much more juice!

Why:

You break up the tissues of the fruit when you roll it on a hard surface, so the juice comes out more easily.

LEMON ICES

Use your lemon juice to make refreshing lemon ices.

- freshly grated lemon peel
- ¾ cup of lemon juice (1 or 2 lemons)
- 1 cup of sugar (or less to taste)
- 4 cups of water
- an ice cube tray without separators, or a metal baking pan
- small paper cups

Grate the peel of one lemon into a small plate or jar.

Simmer the water and the sugar uncovered over medium heat about three minutes until the sugar dissolves. Let it cool and put it in the refrigerator for about an hour until it is cold.

Combine the cold sugar syrup, the grated lemon peel and the lemon juice, and pour it into an ice cube tray. Put the tray in the freezer for about 30 minutes—until ice crystals begin to form. Then stir the mixture well and return the tray to the freezer. Keep stirring every 30 minutes until the mixture is frozen through—about 2 to 2½ hours.

Spoon it into small paper cups—and lick away!

LEMON AS DEODORIZER!

You can clean the smell of fish from your hands by rubbing them with lemon. The smell is caused by nitrogen compounds in the fish. The acid of the lemon changes the nitrogen compounds so that you can rinse them off in cold running water.

Rescuing an Apple

How do you prevent a cut apple from turning brown?

You need:
an apple

lemon juice

What to do:
Cut the apple into quarters.

Let one of the quarters remain on the kitchen table. Place another in the refrigerator.

Sprinkle lemon juice on the other two quarters. Place one of these on the table and the other in the refrigerator.

What happens:
The untreated apple on the table turns brown first. The apple with the lemon juice in the refrigerator stays fresh longest.

Why:
When you cut into apple, you tear its cells, releasing an enzyme called *polyphenoloxidase*. The enzyme speeds up the process by which compounds in the apple (*phenols*) combine with oxygen from the air. This is what produces the brownish pigment that darkens the fruit and makes it taste bad.

The enzyme works more slowly at cold temperatures than at room temperatures. It works even more slowly in an acid like lemon juice, which completely inactivates it.

If you don't have any lemons around, you can use orange juice, but lemon juice is better because it contains more acid.

Not in the Refrigerator

Bananas are picked and shipped green, but green bananas are not digestible. You can ripen them in a few days—but is it true that you should never put bananas in the refrigerator?

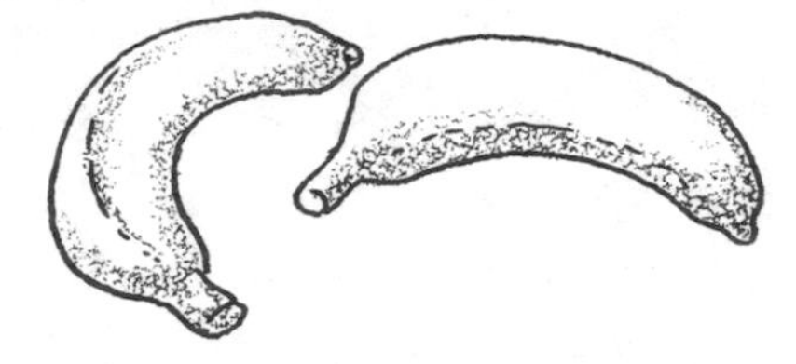

You need:

2 green bananas

What to do:

Place one banana on the counter and one in the refrigerator.

What happens:

Within a few days the banana on the counter turns yellow and its flesh becomes soft and creamy. The one in the refrigerator blackens and its insides remain hard.

Why:

Bananas release ethylene gas naturally to ripen the fruit. On the counter, the skin's green chlorophyll disappears and reveals yellow pigments (*carotenes* and *flavones*). Also, the starch of the banana changes to sugar and the pectin, which holds the cells of the banana firm, breaks down. And so the flesh softens and is easy to digest.

In the cold of the refrigerator, the tropical bananas suffer cell damage and the release of browning and other enzymes. The fruit doesn't ripen and the skin blackens instead of turning yellow.

Once a banana is ripe it is safe to store it in the refrigerator. The skin may darken but the fruit inside will remain tasty for several days.

Powerful Pineapple

Gelatin is a protein that comes from the connective tissue in the hoofs, bones, tendons, ligaments and cartilage of animals. Vegetable gelatin, *agar*, is made from seaweed. Gelatin dissolves in hot water and hardens with cold. We can put all kinds of fruit in it to make terrific desserts—but we're told on the package *not* to add raw pineapple. Why?

You need:

- 1 envelope of unflavored gelatin
- ½ cup of cold water
- a few bits of raw pineapple (or frozen pineapple juice)
- a can of pineapple chunks
- 1½ cups of boiling water

What to do:

Stir gelatin in the cold water and let it stand one or two minutes. Then add the boiling water and stir until all the gelatin is dissolved. Pour into 2 cups or dessert dishes.

To one, add raw pineapple bits or frozen pineapple juice. To the other, add canned pineapple bits or canned juice.

Put both in the refrigerator.

What happens:
The gelatin with the canned pineapple becomes firm. The gelatin with the raw pineapple remains watery.

Why:
Pineapples, like figs and papayas, contain an enzyme that breaks proteins down into small fragments. If you put raw pineapple in gelatin for a dessert or fruit salad, this enzyme digests the gelatin molecules and prevents the gel from becoming solid. It remains liquid.

Cooking stops the enzyme from working. That's why you can add canned pineapple to the gelatin with no bad effects. Since it has been heated, it no longer contains the active enzyme.

PUTTING PINEAPPLE TO WORK

Chefs sometimes simmer raw pineapple with a meat stew to help break down the protein of the meat and make it tender. A bonus is its sweet flavor.

Currying Flavor with a Lime

Sometimes "cooking" starts hours before you light the stove.

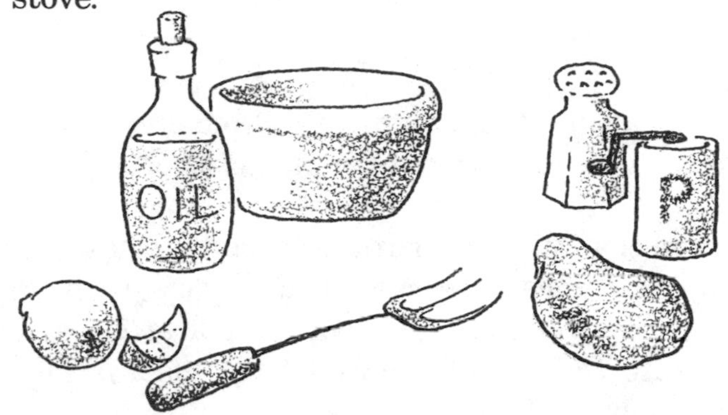

You need:

2 chicken breasts
1 T lime juice
dash of chili powder (optional)
dash of turmeric or cumin (optional)
bowl
2 tsp olive oil
½ tsp minced garlic
1 tsp rosemary (optional)
salt and pepper
wooden spoon

What to do:

Mix the lime juice and olive oil. Add the herbs and stir. Put one of the chicken breasts in a bowl and cover it with the marinade.

Season the other chicken breast with salt and pepper and, if you wish, herbs.

Refrigerate both chicken breasts for an hour or so.

Broil each one. After ten minutes, turn them over and continue to broil for another five to ten minutes. Test them for tenderness and remove them from the oven when a fork goes in easily.

What happens:

The marinated chicken breast cooks faster. The lime not only flavors the chicken but also cuts down on cooking time.

Why:

In marinating, the essential ingredient is an acid such as lime or lemon or vinegar that softens the tissues.

In addition to tenderizing and adding flavor, marinades sometimes preserve color. If you are marinating foods more than an hour or two, it is safer to refrigerate them. Like cooked food, marinating food needs to be refrigerated to prevent the growth of dangerous bacteria.

How to Make Vinegar

Vinegar is often used to flavor salads and tenderize meats. Many vinegars are made from fruit or wine (which is made from fruit). You can use apples to make your vinegar.

You need:
2 apples, cored
blender or juicer
2 jars

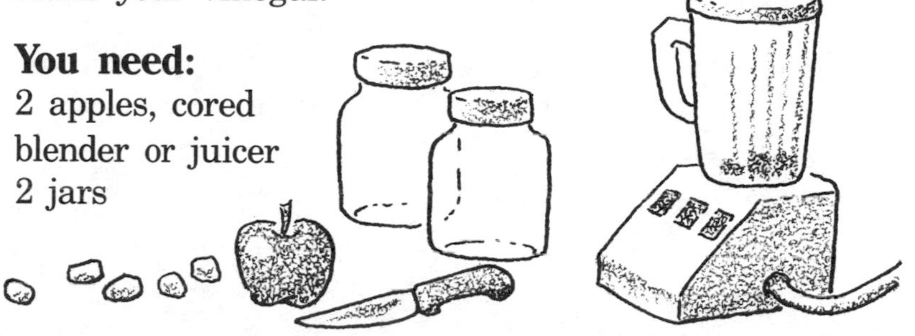

What to do:
Cut the apples into small pieces, place them in your blender or juicer, and press out the juice. Pour half the juice into one jar, and the other half into the other. Place one jar in the refrigerator. Place the other in a warm place.

Compare the color and the odor over a period of a week.

What happens:
Both change, but the juice in the warm place changes much faster—weeks faster! At first you see bubbles and smell alcohol. You may see a thick film forming on top. Then the liquid begins to smell sour.

Why:
Chemical changes have taken place. Yeasts from the skins of the apple and from the air act on the sugars of the apple juice, producing carbon dioxide and alcohol ("hard" cider). Within the week, bacteria in the cider turn it into vinegar.

4. GRAIN: THE STAFF OF LIFE

What happens when you make toast? What puts the bubbles in the pancakes? Why is baking more expensive on a rainy day? What is yeast? And more. . . .

ABOUT GRAIN

Grain, whole or ground into meal or flour, is the principal food of people and domestic animals. Even Goldilocks and the three bears ate cereal, though they called it porridge.

Cereals, breads, rolls, muffins, buns, bagels, pancakes, waffles, spaghetti, macaroni, rice, bulgar, kasha, cookies, crackers, cake—all are grain products. They are made from wheat, buckwheat, rice, rye, oats, maize, barley, and, in Africa and India and China, millet or sorghum. Less familiar grains include *amaranth*, which fed the Aztecs; *quinoa*, a staple food of the ancient Incas; and *triticale*, which modern scientists developed by crossing rye and wheat.

What Is Toast?

Toast is defined in the dictionary as a slice of bread browned on both sides by heat. But what causes the bread to brown?

You need:

a toaster, an electric broiler or an oven

2 slices of bread

What to do:

Place the pieces of bread in the toaster, in the oven under the broiler. Let one stay in twice as long as the other.

What happens:

One turns golden brown. The one kept in too long turns black.

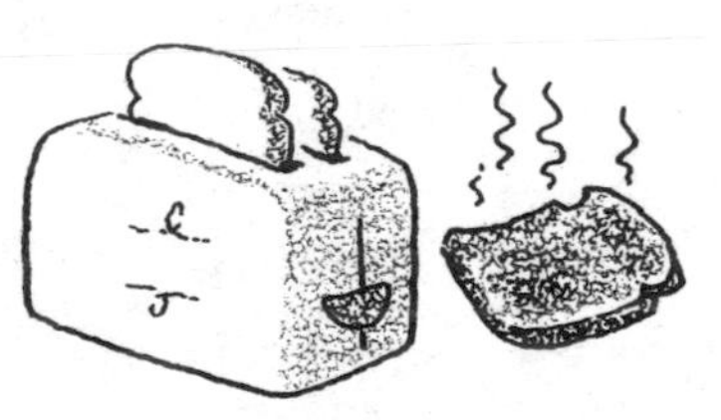

Why:

Too much heat releases the carbon of the starch and sugar. It is this carbon that makes the bread turn black.

Toasting is a chemical process that alters the structure of the surface sugars, starches and the proteins of the bread slice. The sugars become fibre. The amino acids that are the building blocks of the protein break down and lose some of their nutritional value. Toast, therefore, has more fibre and less protein than the bread from which it is made. Some nutritionists believe that when you eat toast instead of bread you are getting color and flavor at the expense of nutrition.

Science for Breakfast

Hot cereal feels good, especially on a cold morning. Does it matter whether you start it in cold or boiling water?

You need:

1½ cup water
salt (optional)
⅔ cup oatmeal
2 small pots

What to do:

Stir ⅓ cup of the oatmeal into a pot with ¾ cup of the water. Bring it to a boil, lower the flame, and simmer for five minutes, stirring occasionally. Cover the pot and remove it from the heat. Let the mixture stand.

In a second pot, bring the rest of the water to a boil. Add salt and pour in the other half of the oatmeal. Lower the flame and simmer for five minutes, stirring occasionally. Again, cover, remove from the heat and let the mixture stand for a few minutes.

Taste the first pot of oatmeal. Then taste the second.

What happens:

Both are cooked and taste good. The oatmeal that started in the cold water is creamier than the oatmeal that started in boiling water.

Why:

As you heat the grains, the starch granules absorb water molecules, swell and soften. Then the nutrients inside are released and are more easily absorbed by the body.

When you start cooking the oatmeal in cold water, the granules have a longer time to absorb the water. The activity starts at 140°F (60°C), well below the 212°F (100°C) boiling point. The complex carbohydrates (*amylose* and *amylopectin*) that make up the starch change. They break up some of the bonds between the atoms of the same molecule and form new bonds between atoms of different molecules. The water molecules then get trapped in the starch granules, which become bulky and eventually break, releasing the nutrients inside.

Add milk and raisins or bananas or blueberries to the oatmeal and you have a terrific dish that also furnishes vitamins, minerals and complete protein.

Why Not Eat Flour Raw?

Flour is the finely ground meal of wheat or other cereal grains. But we can't eat it raw.

You need:

1 T of sugar
1 T of flour
2 glasses half full of cold water

What to do:

Stir the sugar into one of the glasses of cold water.

Stir the flour into the other glass of cold water.

What happens:

The sugar disappears. The flour does not.

Why:

The sugar dissolves in the water. The grains of flour are too big to dissolve. When you stir the flour and water together, you get a paste in which each grain is hanging suspended in the water.

When you chew your food—but before you swallow it—saliva works to help you digest it. Sugar molecules separate and mix with saliva immediately.

Flour, however, is suspended in the saliva, just as it is in the glass of water. The tough wall of plant cells around each grain of flour prevents the starch molecules from getting out. Nothing can get in unless the wall is broken—neither water needed to soften the starch nor the enzymes that would digest it. Heat breaks that wall. That's why flour needs to be cooked.

Popping Popcorn

Making popcorn gives you a good idea of how heat bursts the starch wall. Puffed cereals are made in a similar way.

You need:

about ½ cup of popcorn kernels

¼ cup vegetable oil

a deep, heavy saucepan with a wooden handle and a cover

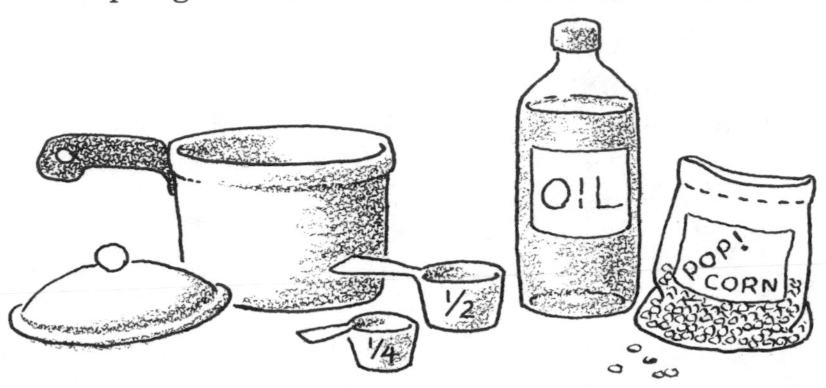

What to do:

Heat the pan on high heat for one or two minutes, and then pour in enough oil to cover the bottom of it. Lower the heat to medium. Add a few kernels and cover the pot. When you hear those kernels start to pop, add just enough popcorn to cover the bottom of pot. Lower the heat. Put on the lid.

Shake the pot from time to time, but *don't* remove the lid while you can hear crackling sounds. When the sounds stop—within a minute or two—remove the pot from the stovetop and uncover it.

What happens:

You have a mountain of popped corn!

Why:
The moist and pulpy heart of the corn kernel is surrounded by a hard starch shell. When the kernel is heated, the moisture in the kernel turns to steam; the heart gets bigger—and the shell bursts.

Grains of starch behave a lot like kernels of corn. When heat breaks the wall, the starch comes out and mixes with water. It is then in a form that we can digest.

All recipes that have flour as an ingredient—cake, biscuits, bread, gravies, sauces, puddings—must be cooked so that the starch in the flour can be released.

BUTTERED POPCORN

Many people eat popcorn exactly as it comes from the pot. But you may want to flavor it with a bit of melted butter or margarine and a pinch or two of salt.

Be sure to wait until after the kernels have popped to add the salt. Doing it beforehand makes the popcorn tough, just as adding salt before they cook toughens lentils and beans. (See page 60.)

Gluten: The Sticky Story

Most flour in English-speaking countries is made from wheat.

There are two types of wheat—hard winter wheat and soft spring wheat. Soft wheat has more starch in it. It is made into soft, powdery cake flours and products that are meant to be tender and crumbly. Hard wheat, with more protein and less starch, is more gritty and coarse. It is better for bread baking because it forms a strong gluten.

What is gluten?

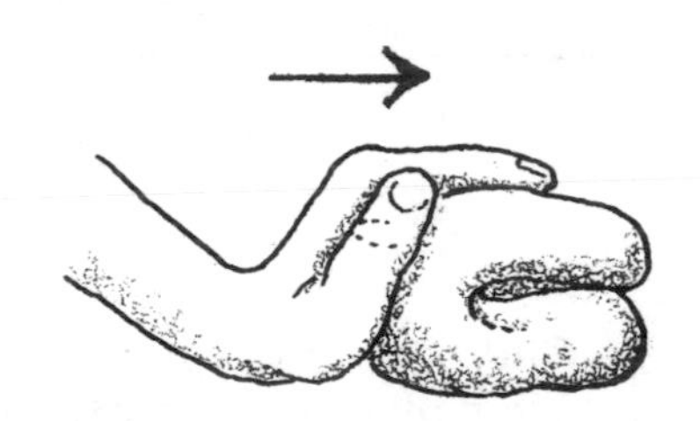

You need:

2 T of warm water
4 T all-purpose flour

What to do:

Mix the flour and water. Roll it into a ball and soak the ball in cold water for 30 minutes. Gently fold and squeeze the dough under running water. Then knead the dough. The illustrations on page 98 and 99 show how.

What happens:

The dough becomes a sticky substance that stretches.

Why:

All-purpose flour is a blend of both soft and hard flour. When you soak the dough in cold water, you wash away the starch, leaving the proteins. When you knead the dough, these proteins (*gliadin* and *glutenin*) interact to form an elastic substance—gluten.

When bread bakes, tiny bubbles of air get trapped inside the gluten. They make the dough rise a little. Without gluten, there would be no raised bread.

Only wheat produces the gluten that traps these air bubbles. That's why most bread and muffin recipes call for some wheat—even rye bread and corn muffins.

To form more bubbles within the gluten and get the dough to rise higher, bakers sometimes add other ingredients, such as yeast (see pages 95–99) or baking soda or baking powder (see pages 102–106).

CONTENTS OF FLOUR

	% Protein	% (Starch)
cake flour	7.3	79.4
semolina flour	12.3	73.5
all-purpose flour	10.5	76.1

Flour also contains 1% fat. The rest is moisture.

Whole wheat flour, graham flour, and cracked wheat all use the whole kernel—the bran (the hard, brown outer cover), the endosperm (the interior food for the germ) and the central germ (the part that sprouts). "White" flour has been refined—the brown bran and the germ have been removed.

WHEAT KERNEL

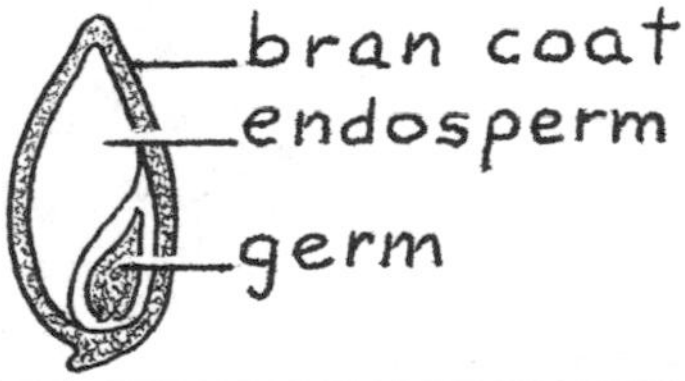

bran coat
endosperm
germ

STORING BREAD

It's best to store bread in a breadbox at room temperature or in the freezer—but not in the refrigerator. Bread gets stale because the water from the interior flows to the crust where it is absorbed by the gluten (see page 90) and starch of the wheat. If bread is left uncovered, its water is lost in the air and the bread gets stale very fast. This happens fastest at temperatures like those of the refrigerator, just above freezing.

Because most of the water remains in the loaf, you can often "revive" stale bread by heating it. However, when bread becomes moldy, it is unsafe to eat.

Popovers: Gluten in Action

The way you mix batter or dough influences the amount of gluten that you develop and whether the baked goods will be spongy or flaky, coarse or fine, tender or tough.

You need:

1 cup (112 g) flour
2 eggs
1 T (15 ml) oil
1 cup (240 ml) milk
½ tsp salt (optional)
muffin tins

What to do:

Preheat oven to 450°F (232°C).

Beat the eggs, add the milk and gradually stir them into the flour to make a smooth batter. Beat the mixture thoroughly with an egg beater or in a mixer. The batter will be thin—like heavy cream.

Grease the muffin tins. Fill them ½ to ⅔ full. Bake at 450° (230°C) for 15 minutes. Then reduce the temperature to 350° (175°C) and bake for 20 minutes more. Don't open the oven door before the time is up.

What happens:

The batter has baked into six to eight crisp, nearly hollow delicious shells!

Why:

When you started, you beat the dough hard to develop the gluten. In the oven, the combination of hot air and steam—formed from the large amount of liquid in the batter—causes the mixture to swell.

If you had opened the oven door, hot air would have escaped—and the popovers would have collapsed.

Hidden Sugar

Would it surprise you to find out that your body converts starch to sugar?

You need:

a pinch of cornstarch or a small cracker

What to do:

Place the cornstarch or the cracker on the tip of your tongue. Mix it with saliva and let it stay for a while.

What happens:

At first, it doesn't taste sweet on the tip of your tongue. But when it mixes with saliva, it becomes quite sweet.

Why:

A molecule of starch is composed of a chain of sugar molecules, which can be broken into separate links by enzymes, proteins that cause chemical reactions. The enzyme (*ptyalin*) in your saliva splits the starch into its sugar links. The links dissolve in your digestive juices and then move easily into your intestines, through the wall into the bloodstream, and then along the bloodstream to the cells.

The Sugar Eater

Yeast, a tiny colorless plant, has been used for thousands of years to put air into breads and cakes.

You need:

warm water
3 glasses
2 T sugar (brown or white) or honey or molasses
2 T flour
a pkg of dry yeast or a cake of compressed yeast
adhesive tape and a felt-tipped pen

What to do:

Pour ⅔ of a cup of warm water into each of the glasses. Number the glasses. Add the sugar to #1. Add the flour to #2. Don't add anything to #3. Next, add an equal amount of yeast to each glass. Let them stand. Observe them after 10 minutes, 20 minutes, 30 minutes. Note the differences.

What happens:

Glass #1 produces bubbles in the first ten minutes.

Glass #2 produces bubbles after 15 or 20 minutes or so.

Glass #3 never bubbles.

Why:

Yeast is a fungus. It feeds on sugar and breathes out carbon dioxide.

In the glass with sugar, it eats quickly—and soon produces carbon dioxide, which makes bubbles. In the glass with the flour, it takes longer because the enzymes in the yeast have to turn part of the flour's starch into sugar before the yeast can digest it.

In the glass of plain water, the yeast has no sugar on which to feed and so does not produce carbon dioxide. Some sugar is required for yeast to feed on.

Too much sugar, however, slows the production of carbon dioxide or even stops activity completely.

Alice's Magic Pill

In Wonderland, Alice ate a little pill and grew and grew and grew. Do you suppose it was made of yeast?

You need:

¼ cup (60 ml) of warm water
1 T of sugar
½ package of dry yeast
2 bowls
3½ cups (392 g) of flour
a measuring cup
¾ cup (180 ml) of water
pinch of salt
wax paper or plastic wrap

What to do:

Put the warm water in a cup. Test a drop of it on your wrist. It should not feel hot. Stir in the sugar and the yeast. Let it stand. Within five or ten minutes, bubbles appear on the surface.

While you are waiting, mix the flour and salt and ¾ cup (180 ml) of water in one of the bowls. Divide the mixture in two, and put half into each bowl.

Add the bubbly yeast to the first bowl. Cover both bowls with wax paper or plastic wrap and let them stand in a warm place from 45 minutes to an hour.

What happens:

The mixture with the yeast doubles in size. The other remains the same size.

Why:

The yeast converts the flour into sugar molecules. It eats the sugar, digests it and uses it for energy, producing carbon dioxide, bubbles which puff up the mixture.

Just Right

Some like it hot, some like it cold, but yeast likes it just right!

You need:

⅔ cup (160 ml) of warm water
2 T sugar
1 package of dry yeast
3 mixing bowls (or cereal dishes)
plastic chopping board
a felt-tipped pen, paper, scotch tape
plastic wrap
3½ cups (392 g) of flour
2 T (30 ml) of oil (olive or corn oil)
5 oz (150 ml) of water
pinch of salt
1 tsp of oil (to oil bowls)
a wooden spoon

What to do:

Dissolve the sugar and yeast in ⅔ cup (160 ml) of water that is warm to the touch. Let the mixture stand.

Mix flour, salt, oil and 5 ounces (150 ml) of water.

When bubbles appear on the surface of the yeast mixture, add it to the flour and mix well with a wooden spoon, or in a mixer. Knead the dough on the chopping board for five to 10 minutes following the illustrations.

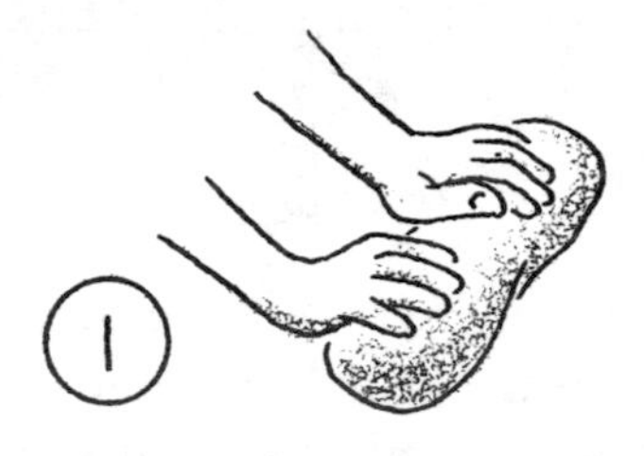

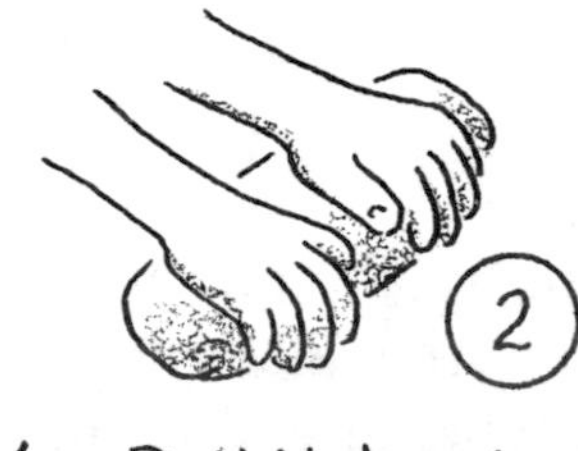

Sprinkle on more flour if the dough gets too sticky to handle. Keep pushing against the ball of dough, pressing into it and turning it to knead it on all sides.

When the dough feels satiny, make it into a ball and divide it in three equal parts. Place each part in an oiled bowl covered with plastic wrap.

Number the samples. Place #1 in a warm place (without a draft). Place #2 in the warmest part of the refrigerator. Place #3 in a hot place—over a radiator or in a hot oven.

Let them stand. Observe them after the first hour and then after several hours.

What happens:

Within 45 minutes to an hour, the dough in a warm place doubles in size. The dough in the refrigerator eventually rises, too, but it takes much longer. The dough in the hot place does not rise at all.

Why:

Yeast requires a moist, warm temperature—above 50° and below 130°F (10°–54°C). Below 50°F (10°C), it is relatively inactive, and above 130°F (54°C), it dies of too much heat.

The Pizza Test

Yeast consists of tiny living cells that make carbon dioxide as they breathe. Its bubbles not only puff up bread and cakes but also pizza. With the dough from the last experiment, you can see just what a difference yeast makes!

You need:

dough #1 from the previous experiment
dough #3 from the previous experiment
2 cookie tins or pie plates
a small can of tomato sauce
a pinch of oregano or
a rolling pin or glass
2 to 4 oz (56–112 g) of cheese (mozzarella, Parmesan or cheddar)
1 to 2 tsp olive oil

What to do:

On a lightly floured board, punch the raised dough with your fists. Knead it for a few minutes and stretch it out or roll it with a rolling pin or the side of a glass to a circle six to eight inches (15–20 cm) in diameter and ¼-inch (.62 cm) thick. Leave the edges a little thicker so they make a rim.

Put the rolled-out dough on an oiled cookie sheet or a pie plate. Let it rise for another 15 minutes.

Roll out the other piece of dough, the one that didn't rise because the yeast was killed. Knead it and then roll it out into the same kind of circle.

Preheat the oven to 450°F (230°C).

Put a layer of cubed or grated cheese on each of the dough circles. Stir a pinch of marjoram or oregano into the tomato sauce. Then pour half of the tomato sauce into the center of each pizza and spread the sauce in circles toward the rim. Top each with another layer of cheese. Place the pie plates near the bottom of the preheated oven. Bake each pizza for about 20 to 30 minutes or until its crust is brown.

Be sure to use a potholder when you take the pizzas out of the oven. Let them stand for about five minutes before you cut them.

Taste each one.

What happens:

The pizza made from the raised dough puffs up even more and has a light, moist taste. The dough of the other "pizza" is unpizza-like—flatter, heavier, and not very tasty.

Why:

The yeast in the raised dough is still active and continues its action during your kneading and for part of the time that the pizza is baking. The other dough bakes as though no yeast had been added.

CHEMICAL BUBBLES

It wasn't until the middle of the 1800s that people started using chemicals to put air into breads and cakes. Today, instead of yeast, we often use either baking soda or baking powder—sometimes both. It takes much less time to bake with them. Batters, such as those used for pancakes and certain cakes, contain much more liquid than doughs used for breads and other cakes made with yeast. These batters are so thin that slow-acting yeast can't trap enough air to make bubbles. That's why we use the modern chemicals.

About Baking Soda

Baking soda is *sodium bicarbonate*—sometimes called bicarbonate of soda. Some people use it for brushing their teeth, for absorbing refrigerator odors or as an antacid for indigestion!

But we can also use baking soda to puff up bread and cake.

You need:

2 tsp of baking soda
a glass of orange juice
or lemonade
a glass of water

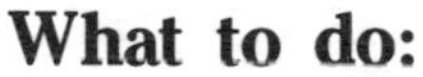

What to do:

Add 1 teaspoon of baking soda to the glass of water.
Add 1 teaspoon of baking soda to the orange juice.

What happens:

Nothing happens in the glass of water.

In the glass with the orange juice, you get bubbles. You have made orange soda!

Why:

When you add an acid (orange juice) to the baking soda, you free the carbon dioxide of the baking soda—the bubbly gas.

Try adding baking soda to buttermilk, sour cream, yogurt, molasses, apple cider. They are all acidic and they will all bubble.

When baking soda is added to dough made with any of these or other acidic liquids, bubbles form and cause the dough to rise.

About Baking Powder

How is baking powder different from baking soda?

You need:

2 glasses of water
½ tsp baking soda
½ tsp baking powder

What to do:

Add the baking powder to one glass of water. Add the baking soda to the other.

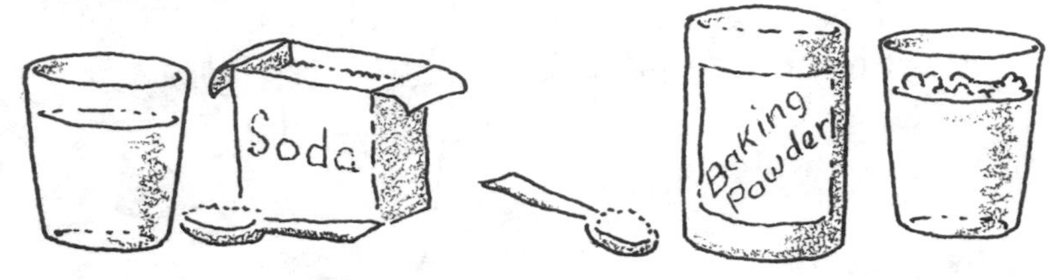

What happens:

The water with the baking powder bubbles. The water with baking soda does not.

Why:

Baking soda is an alkali, the chemical opposite of an acid. When it combines with an acid, it forms carbon dioxide.

Baking powder is a combination of baking soda and an acid. When you add baking powder to water or milk, the alkali and the acid react with one another and produce carbon dioxide—the bubbles.

There are three types of baking powder. Each one contains baking soda. In addition, they each contain an acid—either cream of tartar (tartrate baking powder), monocalcium phosphate (phosphate baking powder) or a combination of calcium acid phosphate and sodium aluminum sulfate (double-acting baking powder). More about these on page 105.

Powder Versus Soda

What happens if we add baking powder to an acid?

You need:

2 half-filled glasses of sour milk or orange or lemon juice
½ tsp baking powder
½ tsp baking soda

What to do:

Add baking powder to one of the half-filled glasses of sour milk and baking soda to the other glass.

What happens:

The sour milk with the baking powder does not bubble as much as the one with the baking soda.

Why:

When you add baking powder to an acid, you are tampering with the balance of acid and alkali. You are adding more acid than alkali. The result is that you actually reduce the amount of carbon dioxide produced.

Therefore, if you want to bake with sour milk *or* buttermilk instead of regular milk, you could do it by eliminating the extra acid. You would just replace each teaspoon of baking powder in the recipe with ½ teaspoon of baking soda.

Trapped Bubbles: Pancakes

A delectable way to see bubbles trapped in dough is to make pancakes. It's also a good way to use up milk that has turned sour.

griddle or pan
wax paper with oil on it
½ cup (56 g) all-purpose flour
1 tsp sugar
pinch of salt (optional)
⅓ tsp baking soda
1 egg white
⅓ cup (80 ml) sour milk or buttermilk
1 tsp oil
spatula (pancake turner)

Wipe a griddle or pan with the oiled wax paper and place it on the stove. Warm it over medium heat.

Stir the dry ingredients together. Beat the egg, add the sour milk and oil. Add the liquid to the dry ingredients gradually, stirring only until the batter is smooth. Don't keep stirring once it is smooth or the pancakes will be tough.

When a drop of water sizzles on the heated griddle, drop two heaping tablespoons of the mixture onto the hot greased griddle. Cook each pancake until the top is full of bubbles and the underside is brown. Then flop the pancake over and brown it quickly on the other side. Makes 6 pancakes.

Serve with maple syrup, honey or sugar, and butter or margarine.

If you prefer pancakes made with regular milk, substitute ⅔ tsp of baking powder for the baking soda. (See page 105.)

Model Muffins

If you want light, fluffy muffins, take care to treat the batter right! See what happens if you don't!

You need:

1 cup (112 g) all-purpose unbleached flour
a small egg
3 T sugar
1 tsp baking powder
¼ tsp ground cinnamon
¼ tsp ground nutmeg
2 large bowls
a wooden spoon
¼ cup (60 ml) oil
½ cup (120 ml) milk
a whisk (optional)
1 small bowl
3 muffin pans
oil or margarine for greasing

What to do:

Grease the muffin pans.

Using the smaller bowl, beat the egg with a spoon or a whisk. Then add the milk and oil.

In one of the large bowls, combine the flour, sugar, baking powder, cinnamon and nutmeg. Make a hole in the center of the dry ingredients. Dump the liquid ingredients into the hole. Stir the mixture about 12 to 14 times, just enough to moisten the dry ingredients. The batter should be rough and lumpy.

Pour half of the batter into a second large bowl. Mix that batter until it is smooth.

Spoon a heaping tablespoon of the lumpy batter into one of the muffin cups so that it is ⅔ full. At the opposite end of the same muffin pan, do the same thing with the smooth batter.

Repeat the process with the two other pans. Now you have three muffin pans, each with one muffin of smooth batter and one of lumpy batter.

Turn on the oven to 400°F (205°C). Don't preheat it; instead, immediately put in one muffin pan.

After 10 minutes, put in the second pan.

In about 25 to 30 minutes, when the muffins are a golden brown, remove them from the oven. (Use pot holders!)

Then, put the third muffin pan into the hot oven and turn the heat up to 450°F (230°C). After 25 or 30 minutes, remove this pan from the oven.

Let them all cool and sample each muffin.

What happens:

The muffins from the lumpy batter in the preheated oven—pan #2—look and taste the best.

Why:

For delicious muffins, you don't need to work hard! Overmixing develops the gluten and results in knobs or peaks on the top and long holes or tunnels inside the muffins.

It is also important to preheat the oven before you put the muffins in. If the oven is not hot enough, the muffins will be flat and heavy. That's because the baking soda isn't activated soon enough to cause the batter to rise.

However, if the oven is too hot, the carbon dioxide goes to work too soon, and muffins will be poorly shaped and tough.

Weather and Cookies

Make the following cookies on two different days—one sunny and dry, the other rainy. The cookies will taste good on both days but . . .

You need:

½ cup of butter or margarine
¼ cup of sugar
½ tsp vanilla extract
wooden spoon
1 cup (112 g) of flour plus 1 to 3 extra T (8–24 g)
a mixing bowl
a cookie sheet

What to do:

Let the margarine stand at room temperature for 10 to 15 minutes. Then put it in a mixing bowl and add the sugar and vanilla extract. Cream the mixture well with a wooden spoon or in a food processor or electric mixer. Add the flour and continue mixing. When the dough is thoroughly mixed and smooth, remove it from the bowl and form a ball. If it is sticky, roll it in flour until it feels satiny.

Wrap the dough in wax paper and refrigerate for an hour or more.

Preheat the oven to 325°F (165°C).

Cut the ball in half and roll out two logs, adding flour if the dough is sticky.

Slice thin, as in the illustration, and place the circles a half-inch apart on an ungreased cookie sheet. Bake in center of the oven for 20 minutes or until the bottoms of the cookies are slightly brown.

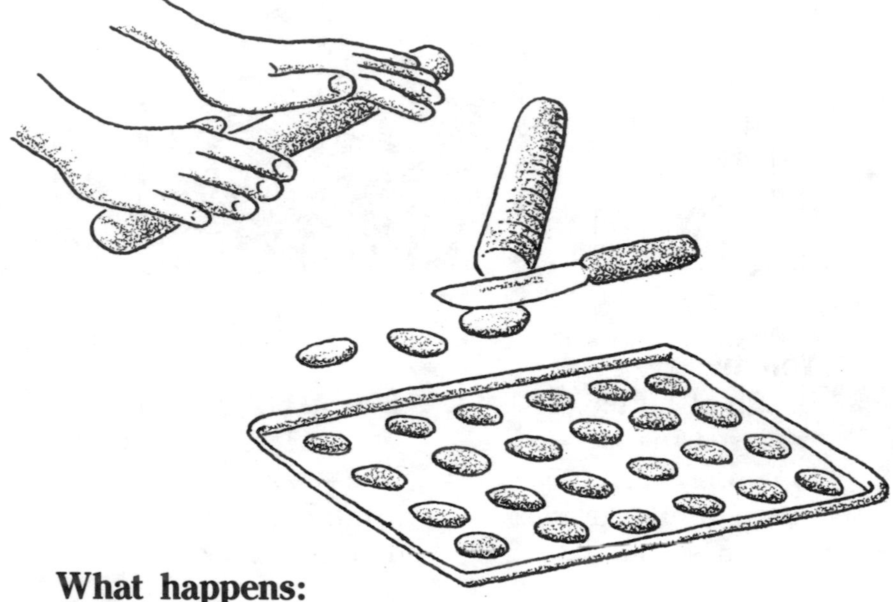

What happens:

On both the rainy day and the sunny day, you'll end up with 4 to 5 dozen great cookies. But it takes several more tablespoons of flour on the rainy day than it does on a dry, sunny day!

Why:

On a rainy day, the dough soaks up water from the air, gets sticky and is harder to handle. You therefore have to use more flour than on a dry day.

WHAT "WHERE" HAS TO DO WITH IT

If you live in an area where the altitude is 500 feet or more above sea level (166 m), any dough with yeast or baking powder or soda will rise more quickly than it would at sea level. This is because the blanket of air (atmospheric pressure) is lighter. The carbon dioxide encounters less resistance from the surrounding air, and so it rises higher—with more force and more rapidly. This may make for tough, tasteless baked goods. The solution: Use less yeast or baking powder than you would at sea level.

Some prepared commercial mixes solve the problem by directing you to add more flour at altitudes of 3,500 feet or more.

5. MAKING FOOD LAST

How do salt and sugar preserve food? How do you change a grape into a raisin? Why freeze herbs? What is cheese? And more about making food last.

ABOUT PRESERVATION

To make foods last longer, we try to halt the growth of the bacteria, fungi and moulds that make it spoil. Sometimes we need to destroy destructive enzymes or prevent oxidation. Either of these processes changes the color, texture, taste or nutritional value of a food.

We use many methods: drying, salting, smoking, pickling, canning, refrigerating and freezing, among others.

For centuries, particularly in warm countries, people used salt to preserve many types of meat and vegetables. Beef still comes to the table as corned beef, cabbage as sauerkraut, cucumbers as pickles. Salt draws moisture from food by osmosis (see page 44) or absorption. This discourages the growth of bacteria. In dry curing (or corning), food is buried in salt. Other techniques involve soaking food in brine, a salt-and-water solution, or injecting salt into the food.

Smoking meat and poultry and fish is also an age-old practice. The food is hung, usually in a special smokehouse, above hickory, apple, maple or other aromatic wood chips burning at a low temperature—sometimes for days. The longer the smoking process, the stronger the flavor and the longer the food lasts.

When we talk about pickling, we usually think of pickled cucumbers. Cucumbers turn into pickles with the help of vinegars and other acids, spices such as dill, and salt and sugar. But other vegetables and fruit as well as fish and meats can also be preserved by pickling.

Canning dates back only to the early 1800s. It involves the rapid heating of sealed sterilized containers, glass or tin. Salt is often added and so is sugar. Like salt, sugar acts as a preservative, helping to keep mold from growing on such foods as jams and jellies.

In addition to these methods, commercial food manufacturers also add chemicals to extend the time food can remain on store shelves.

Hocus-Pocus—Raisins

A raisin starts life as a grape and a prune as a plum. What happens to these juicy fruits? See for yourself.

You need:

a bunch of white seedless grapes
2 drying trays (make your own by spreading cheesecloth or wire mesh over old frames or kitchen cooling racks)
a pot of boiling water
a strainer
4 empty juice cans or large stones

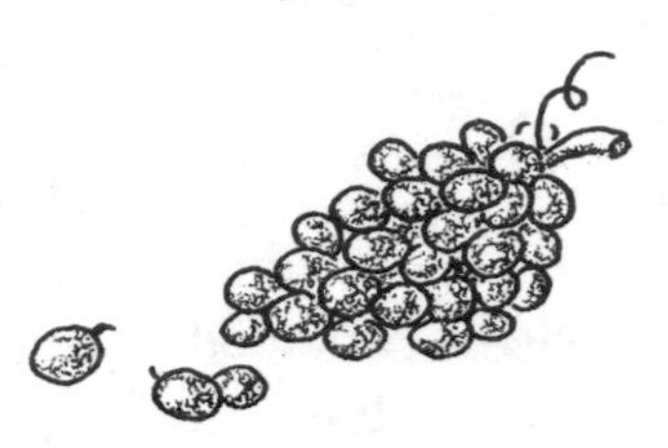

What to do:

Wash the grapes in cold water and remove those that are bruised. Pull out the stems and place the cleaned grapes in a strainer. Dip them into a pot of boiling water so that the skins break.

Spread the grapes on one of the drying trays so that they don't touch. Using the empty cans, prop the second tray over the first.

There are two methods you can use.

1. For four or five days, place the trays by a sunny window, turning them every hour so that the fruit dries evenly.
2. Or place the trays on the middle rack of a pre-heated oven (140°F) (60°C) and let them remain overnight.

When you think they may be dry, remove one or two of the grapes. Let them cool, and test them for moisture. If they still have water in them, let the fruit dry for another hour or so. Then test. If the grape is pliable and chewy, remove the rest from the drying tray.

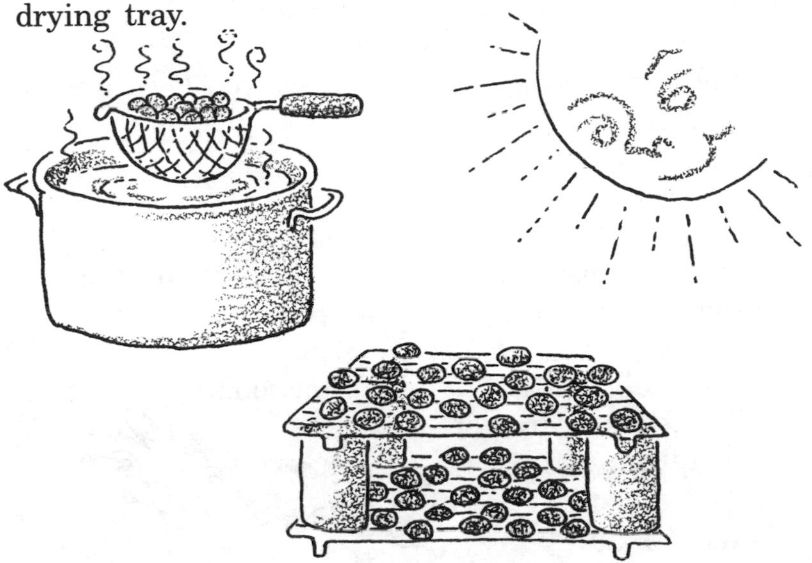

What happens:

You have raisins!

Put them in a plastic bag and they will last for months and months.

Why:

Drying as a means of preservation is thousands of years old. Normally fruit rots in a week or less at room temperature. Even in the refrigerator it will rot after a few weeks.

Fungi that start as spores—tiny seedlike cells—drop from the air and feed on the fruit's sugars and starches. When you dry out the grapes, you are taking away the moisture that the fungi need in order to grow. As long as the dried fruit can't take in moisture from the air, it will stay edible for many months.

Freezing Herbs

Certain herbs, such as parsley, chervil and chives are preserved better by freezing than by drying. The secret is to package them so that you keep air out and moisture in.

You need:

a bunch of herbs (parsley, basil, dill, sage, thyme or chives)
thick small plastic bags, jars or freezer containers
paper towels
labels and felt-tipped pen
drinking straw (optional)

What to do:

Wash the herbs in cold water and remove any leaves that are rotting. Drain and pat them dry with paper towels.

Strip the leafy herbs from their stems. Package the leaves in small plastic bags leaving ½ inch (1.25 cm) of headroom. You can use a drinking straw to remove as much air as possible. Or making sure that no water enters the bag, you can dip it in a pot of water. This pushes the plastic against the food, forcing out all the air.

Seal the bag tightly, using freezer tape if the bags are not self-sealing. Label each one with the name of the herb and the date, and place the bags in the freezer, preferably at 0°F (−18°C).

What happens:

At that low temperature, herbs last up to a year and maintain flavor, color and nutrients. You can add them to soups, stews, sauces, salads and other foods while they're still frozen.

Why:

Enzymes, protein molecules that speed chemical reactions, harm foods by changing their color, texture, taste and nutritional value. Like heating, freezing slows down active enzymes and delays the spoiling process.

You remove the air because air pockets between the food and the plastic bag collect moisture from the food, which results in frost and freezer burns. As ice crystals form, the water expands and ruptures cell membranes and walls.

Because food expands during freezing, you don't fill the bag completely. The bag or other container will split if it's too full for the contents to expand freely.

If the freezer temperature is higher than 0°F (−18°C), the herbs will not keep their flavor as long. Each 10°F (−12°C) above zero cuts the storage life in half!

You can substitute frozen herbs for fresh in recipes, but remember to use them while they're still frozen. If you let them thaw, microbes and enzymes have time to wilt and darken them.

ABOUT HERBS AND SPICES

For hundreds of years, herbs and spices were a symbol of wealth, valued because they preserved food or disguised its smell when it was spoiled. They also served as medicines.

Spices are the dried flavors made from the buds, flowers, fruit, bark and roots of fragrant tropical plants. Herbs are the leaves, stems or flowers of aromatic succulents grown in temperate climates.

- ★ Fresh herbs will keep for a week in the refrigerator. Wrap them in paper towels inside a plastic bag.
- ★ Cut, crush or mince the herbs just before you use them.
- ★ Buy dried herbs and spices in the smallest quantity possible—they lose flavor with age and exposure to air.
- ★ Use less dried herb than fresh—⅓ to ½ teaspoon of dried to a tablespoon of fresh.
- ★ Presoak dry herbs for a few minutes in lemon juice, soup stock or oil for more flavor.
- ★ Add herbs the last 10 or 15 minutes of cooking.
- ★ If you've added too much of an herb, add a raw potato to the cooking pot. It will take up some of the excess flavor and save your dish from being too spicy.
- ★ Put a bay leaf in your flour canister to help protect against insects. Bay leaves are natural insect repellents.

To Freeze or Not to Freeze

Is freezing a good way to preserve any food?

You need:

lettuce leaves or green pepper or tomato
cloves of garlic or onion
2 T (30 ml) cottage cheese
3 plastic bags
labels and freezer tape

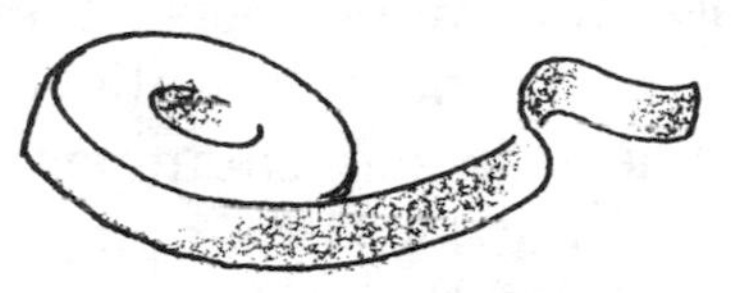

What to do:

Cull the vegetables, wash and pat them dry. Put them each in a plastic bag, whole or cut up, and remove the air as described on p. 116. Seal the bags, label them and place them in the freezer. Spoon the cottage cheese into a plastic bag. Remove the air, seal the bag, label it, and also place it in the freezer.

After two or three days, remove all the bags from the freezer and thaw the contents.

What happens:
The foods are no longer appetizing. The lettuce and tomatoes have lost their crispness and become limp. The garlic has become stronger. The cottage cheese has separated and become grainy.

Why:
When the water cools, it expands and turns to ice, damaging the cell walls of the foods. This loss of crispness is not so important if the food is to be cooked, but foods that we eat raw, like lettuce, tomatoes, and cottage cheese, definitely lose their appeal.

Salted foods also don't freeze as well as unsalted foods. This is because the salt lowers the freezing point and gives the enzymes more time to work.

Preserving a Pear

One great way to preserve fruit is to convert it into jam or jelly.

In the past these have been made with heavy concentrations of sugar and pectin and stored in sterile jars. In fact, for commercial manufacturers to label a product "jam" or "preserve," the U.S. Federal law requires that 65 percent of the final product be sugar. In Europe, there is a similar requirement for products labeled "conserve." A certain amount of sugar and acidity prevents the growth of dangerous microorganisms.

But now, fruit spreads, sweetened with fruit juice concentrates instead of sugar, are being sold commercially. We can easily and safely make these in our kitchen to refrigerate for up to a month or two.

You need:

2 pears
1 tsp grated peel (zest) of lemon
½ small can of thawed frozen apple juice concentrate
or
1 cup of apple juice
½ cup of water
¼ tsp vanilla
1 tsp lemon juice

What to do:

Peel the pears and remove the cores. Then cut the pears into cubes. Put the cubes in a bowl with the grated lemon peel and lemon juice.

Heat the thawed apple juice, water and vanilla for 10 minutes or so. (It's okay to use canned or bottled apple juice, but if you do, don't add water.) Add the pears and lemon juice.

Bring the mixture to a boil. Then lower the heat and, stirring frequently, cook it from 30 to 40 minutes, or until it thickens. Place it in a clean jar and refrigerate.

What happens:

You have pear jam that will keep for a month or two.

Why:

The acid of the lemon juice and the sugar of the apple juice prevent the growth of dangerous microorganisms.

You can make chunky preserves by slicing your pears into eighths. Add grated lemon peel and lemon juice and cook in apple juice for 20 to 30 minutes, or until soft but not mushy.

Little Miss Muffet

Making cheese is one of our oldest ways of preserving milk. How is it done? Was Miss Muffet eating milk or cheese? And did her curds and whey come from a cow, a goat, a sheep, a mare, a camel, a llama, a reindeer or a buffalo? Milk and cheese can come from all these—and other animals!

You can make your own cottage cheese and observe the start of the process by which all cheeses are made.

You need:

a glass of milk
a pinch of salt
rubber band or length of string
a bowl
cheesecloth or a strainer lined with a clean cloth

What to do:

Let the milk stand for two or three days at room temperature until it sours and starts to form chunks. Add a pinch of salt.

Using a rubber band or length of string, fasten a square of cheesecloth over a wide bowl. (Or line a strainer with a clean cloth—an old cotton handkerchief—and suspend the strainer on the lip of the bowl.)

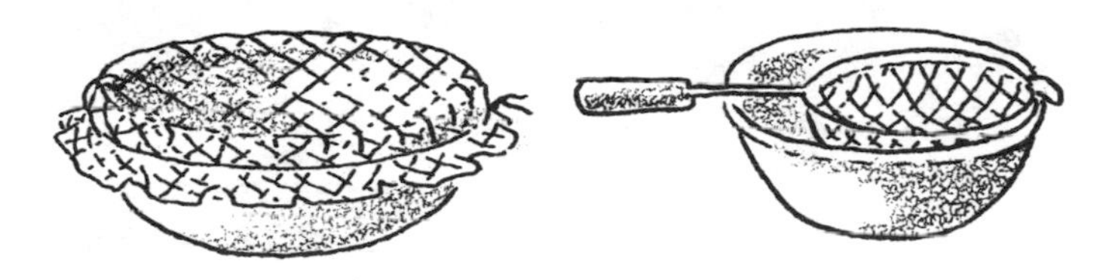

Empty the sour chunks onto the cheesecloth. Let them drip for two or three hours.

What happens:
You have cottage cheese in the cheesecloth.

Why:
Harmless bacteria act on the sugar in the milk to sour it and change the sugar into acid. The acid curdles the milk and separates it into a liquid (whey) and little solid chunks (curds). The curds contain a protein (*casein*), mineral salts, and the butter fat of the milk.

Cottage cheese is a "fresh" cheese and lasts a relatively short time. Many other cheeses are cured—aged and ripened—by adding different molds and bacteria for varying periods of time. These give the cheeses their particular flavors and a longer life.

WHY SWISS CHEESE HAS HOLES

Carbon dioxide makes the holes in the Swiss cheese. It is released by bacteria that are added during the curing process.

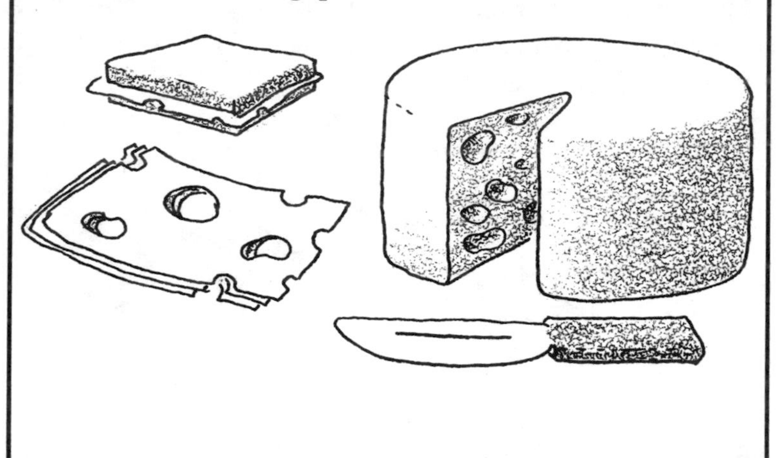

GLOSSARY

acids—a large class of compounds that are capable of neutralizing alkalis and which range from benign sour-tasting citric acids, like lemons, limes and oranges, to hazardous, poisonous sulfuric and hydrochloric acid.

alkalis—also known as bases—any of numerous bitter-tasting, soapy-feeling substances that dissolve in water and neutralize acids to form salts. These include carbonates, like sodium bicarbonate (baking soda) and sodium carbonate (washing soda), and caustic hydroxides like lye, limewater and ammonia, useful commercially and in the home.

amino acids—the building blocks of proteins. The body manufactures all but nine amino acids. These must be obtained from the food we eat. Meat, fish, poultry, dairy products and eggs contain all nine essential amino acids.

baking—dry heat method of cooking, especially in an oven

boiling point—point at which a liquid turns into a gas

broiling—cooking by direct exposure to high heat, over a grill or under an electric element

calorie—a measure of energy. A calorie is a quantity of food capable of producing the heat and energy needed to raise the temperature of one kilogram of water one degree centigrade. An ounce of carbohydrates or protein is equal to 115 calories; an ounce of fat is equal to 255 calories. A person's daily calorie need depends on age, weight, and level of activity.

carbohydrates—the sugars and starches that supply energy help the body use fat efficiently, or provide fiber. They are compounds made of carbon, hydrogen and oxygen, most of which are formed by green plants. Simple carbohydrates are honey, sugar and fruits. Complex carbohydrates are grains and cereals, dried beans, root vegetables and potatoes.

enzyme—protein molecules that break down or build up materials inside the body but are not changed themselves (catalysts). Human digestive enzymes break up proteins into individual amino acids and starch into individual glucose units.

digestion—the changing of food to a form the body can use

fiber—the parts of cereal grains, fruits and vegetables, seeds, legumes and nuts that cannot be digested. Fiber aids digestion and elimination by carrying waste products with it as it leaves the digestive tract and by absorbing fluids that make wastes soft enough for easy passage.

fungus—a plant like mushrooms or yeast that cannot manufacture its own food but lives off decaying organisms around it. This kind of plant is called saprophytic.

minerals—small amounts of minerals such as magnesium, phosphorus, fluorine, potassium, chlorine, copper, iron, iodine, sulfur, and zinc are needed for teeth, bones and health. Larger amounts of calcium and sodium are needed.

molecule—one or more atoms that are the smallest particle of an element or compound that retains the properties of the substance.

organic compound—a group of compounds containing the carbon necessary for life.

proteins—a group of organic compounds containing nitrogen, which our body needs to build and repair tissue, red blood cells and enzymes.

osmosis—the flow of a liquid through a thin membrane from an area of greater concentration to an area of lesser concentration of water

vitamins—special nutrients needed in small quantities but essential to life. A, D, E and K dissolve in fat and can be stored for a long time in the body. Eight B vitamins and vitamin C dissolve in water. Because they are not stored in the body very long, foods providing them must be eaten daily—whole grains, meat or beans for the B's, citrus fruits, melon, berries or leafy green vegetables for the C.

yeast—a group of about 160 species of single-celled microscopic fungi, some of which spoil fruits and vegetables or cause disease. Others are used in making bread and alcoholic drinks.